W9-BTI-956

RED DRAGON RISING

RED DRAGON RISING

*Communist China's
Military Threat to America*

EDWARD TIMPERLAKE
AND
WILLIAM C. TRIPLETT II

Since 1947
**REGNERY
PUBLISHING, INC.**
An Eagle Publishing Company • Washington, DC

Library of Congress Cataloging-in-Publication Data

Timperlake, Edward.
Red dragon rising: communist China's military threat to America /
Edward Timperlake, William C. Triplett II.

p.   cm.
Includes bibliographical references and index.
ISBN 0-89526-258-4 (acid-free paper)
1. China—Military policy. 2. China—Armed Forces. 3. China—Military
relations—United States. 4. United States—Military relations—China.
5. World politics—1989–   I. Triplett, William C. II. Title.
UA835.T497 1999
355'.031'09510973—dc21   99-045384
                                                                           CIP

Published in the United States by
Regnery Publishing, Inc.
An Eagle Publishing Company
One Massachusetts Avenue, NW
Washington, DC 20001

Distributed to the trade by
National Book Network
4720-A Boston Way
Lanham, MD 20706

Printed on acid-free paper
Manufactured in the United States of America

10 9 8 7 6 5 4 3 2 1

Books are available in quantity for promotional or premium use.
Write to Director of Special Sales, Regnery Publishing, Inc.,
One Massachusetts Avenue, NW, Washington, DC 20001,
for information on discounts and terms or call (202) 216-0600.

To the Fallen,
Tiananmen Square,
June 4, 1989

TABLE OF CONTENTS

CAUTION

In order to accurately chronicle the Chinese People's Liberation Army (PLA), we have had to include graphic descriptions and images of PLA atrocities that some readers may find offensive.

—The authors

CHAPTER 1
RED DRAGON RISING

O n a crisp autumn evening in 1997, guests at a White House state dinner listened as President William Jefferson Clinton warmly welcomed President Jiang Zemin, the leader of the People's Republic of China (PRC), the Communist regime that holds a fifth of the world's population in its grasp. That day, the American president had forged a strategic partnership with the Chinese Communist government,[1] a totalitarian regime that has been responsible for enormous evil. Clinton's effusive, almost fawning toast celebrated peace and prosperity between the United States and the PRC.[2]

It would take a year and a half for the House of Representatives, the people's body, to begin stripping away the Clinton administration's camouflage disguising the truth about Communist China. In May of 1999 a select congressional committee, in a unanimous, bipartisan vote, identified President Clinton's betrayal of his most sacred trust—safeguarding the national security of America. This is President Clinton's legacy.

* * *

At last, the report was made public. On May 25, 1999, Representatives Christopher Cox (R-California) and Norm Dicks

(D-Washington) delivered their committee's unanimous findings. After a five-month struggle with the Clinton White House over the report's declassification and release, Cox and Dicks revealed the central conclusion of their committee's six-month investigation: The PRC has stolen America's most advanced nuclear weapons secrets, and Chinese espionage "almost certainly continues today."[3]

The Cox Report was a stunning document. It made headlines across the country and dominated television and radio news programs. For the first time, many Americans began to consider that Communist China—armed with nuclear weapons—might target the United States.

But inevitably, as one news cycle ended, another began, and America's focus shifted away from the Cox Committee's shocking findings.

Still, the larger implications of the Cox Report cannot safely be ignored. In fact, the congressional committee told only a small part of the story. Communist China poses an extraordinary military threat to the United States and the rest of the world.

The thesis of this work is simple: The democratic countries are about to be unpleasantly surprised by the emergence of a hostile, expansionist, nondemocratic superpower armed with the most modern weapons… and it will be our fault.

In short, through a misguided foreign policy that has sacrificed national security for money and personal political power, the Clinton-Gore administration has materially assisted Beijing's military ambitions.

THE PLA: COMMUNIST POWER AT THE POINT OF A BAYONET

For a half-century the Chinese Communist Party (CCP) has ruled China. Its administrative apparatus is known as the People's Republic of China, but essential to the Party's survival is its military arm, the People's Liberation Army (PLA), which enforces the CCP's power.

When one asks, "Is China a threat?" the real question should be, "Is the PLA a threat?" To answer that question, consider that the PLA:

- is targeting the American people with nuclear weapons[4] and is developing entirely new generations of land-, sea-, and space-based strategic weapons systems capable of threatening any location on the planet;
- is preparing for computer warfare ("information warfare") against the American homeland, putting all Americans at risk;
- is selling the critical equipment necessary to make nuclear weapons, poison gas, biological weapons, and ballistic and cruise missiles to the most brutal terrorist regimes—Iran, Iraq, Libya, Syria, and North Korea—posing a direct strategic threat to the United States and its allies, including Israel, Japan, India, and even southern Europe;
- conducted a ballistic missile blockade of Taiwan and intends to end democracy on the island—by force, if necessary;
- murdered thousands of its own people during the June 1989 Tiananmen Square massacre in order to maintain the Chinese Communist regime in power;
- has an unparalleled history of aggression, including, since 1949, unprovoked military attacks on South Korea, India, Vietnam, and the Soviet Union, as well as armed subversion against Malaysia, Singapore, Thailand, the Philippines, Indonesia, and other Asian and African countries;
- has invaded and seized control of one of its neighbors, Tibet, and is slowly taking control of another, Burma;

- is, largely in secret, building a vast war machine based on the most modern weapons and tactics to support the PRC's plans for regional domination;
- is financing its military modernization program by its arms sales to terrorist nations—the PLA and its related industries are willing to sell any weapon to any person, group, or country, no matter what the consequences, to support the silent military buildup; and
- is gathering military technology by spying, colluding with Western high-tech firms, and perhaps gaining influence within some Western governments.

Based on this horrifying record, we believe that the PLA is a very real threat. First and foremost, the PLA, the ultimate enforcer of Communist rule in China, is a threat to its own people, as it demonstrated when PLA troops shot down thousands of innocent Chinese demonstrators at Tiananmen in June 1989. But the PLA also threatens neighboring Asian countries that stand in the way of its drive for dominance, as well as nations around the world that could be victimized by PLA arms sales to terrorist nations. We further believe that unbridled PLA modernization is fast making it a threat to the rest of the world, especially the United States, which must aid those who are threatened.

The Clinton-Gore administration has welcomed to Washington six of the seven uniformed generals directly responsible for the Tiananmen Square massacre.

These threats—to the Chinese people themselves, to all of Asia, and to the United States and democratic countries around the globe—make continued Communist Party domination of China today's leading national security concern.

IGNORING THE RECORD

Our first book, *Year of the Rat*, presented the results of our investigation into the most successful case of high-level bribery in Washington. A hostile foreign power penetrated the American government with money and influence to an extent never before seen in the history of our country. Millions of dollars in illegal campaign contributions originating from Communist Chinese sources flowed into the Clinton-Gore campaign coffers. In return, President Clinton and Vice President Gore made a number of policy decisions that directly promoted their benefactors' ambitions, to the ultimate detriment of the American people. Although the results of our investigation have been confirmed by the Cox Report and subsequent news accounts, we concluded that the consequences of these acts could not be understood properly without an appreciation of Communist China and its military.

Red Dragon Rising is a Bill of Indictment against the Chinese People's Liberation Army and the Clinton-Gore administration for helping the Communist Chinese regime move closer to its dangerous goals.

The Clinton-Gore administration refuses to see Communist China for what it is: a brutal, expansionist regime that will go to any length to fulfill its territorial ambitions. The PLA's fifty years of armed aggression and subversion against China's neighbors have accounted for the deaths of millions and for devastation throughout Asia. And today, the Tiananmen Square killers infest the highest ranks of the Chinese military establishment. Having butchered young Chinese people without hesitation, they would have no reluctance to do the same to foreigners.

But on President Bill Clinton and Vice President Al Gore's watch, the PLA's image has been rehabilitated, despite the atrocities it committed just a decade ago at Tiananmen. The Clinton-Gore administration has gone so far as to welcome to Washington six of the seven uniformed generals[5] directly responsible for the massacre. In one case, the American taxpayers even paid for the Hawaiian vacation of a Tiananmen Square general and his party.[6] Under ordinary circumstances, a foreign army general known to have the blood of his countrymen on his hands cannot safely travel to a democratic country, as General Pinochet of Chile discovered in the winter of 1998–1999, when Britain stripped him of his diplomatic status and arrested him. But if the blood on the hands of the general comes from young Chinese, the result, it seems, is entirely different—at least under Clinton and Gore, as long as they're receiving Chinese campaign contributions.

In the United States, none of this is made public. But photos of Chinese generals smiling with President Clinton in the Oval Office have run on the front page of almost every major Chinese and Hong Kong newspaper, sending a clear message to Chinese patriots: The United States government at its highest levels endorses the Chinese people's oppressors.

Moreover, by seeking to form a "strategic partnership" with the PRC, President Clinton and Vice President Gore have disregarded fundamental national security concerns. Under the Clinton-Gore administration democratic Taiwan has been bullied with nuclear-capable missiles, the Philippines has lost some of its territory, PLA weapons of mass destruction have gone to terrorist nations, and the PLA has developed a computer warfare capability that threatens to turn American cities into infernos.

In documenting the crimes of the Chinese People's Liberation Army, its bloody history, its ambitions, its burgeoning capabilities, and the risks it poses unless Americans and their demo-

cratic allies can stop it, *Red Dragon Rising* will reveal how Bill
Clinton and Al Gore abetted the leading national security threat
of our time, including:

- how secret Oval Office meetings with senior PLA generals rehabilitated the murderers of Chinese young people
  at Tiananmen and discouraged the democracy movement
  in China;
- how the failure to support America's democratic allies in
  Asia makes war in the region more likely;
- how the coverup of Chinese arms sales to terrorist countries puts the entire world at risk; and
- how the presidentially authorized release of six hundred
  American supercomputers at the beginning of the 1996
  presidential campaign led directly to an unprecedented
  threat to America.

If the United States and the other great democracies do not
move quickly to counter Communist China's military ambitions, America and its allies will soon suffer the devastating consequences of having ignored the dragon rising in the East.

PART I
A HOSTILE FORCE

CHAPTER 2
ASSAULT ON A CITY

The PLA conducted a multi-division assault on Tiananmen Square.[1]

—Brigadier General Michael T. Byrnes, defense attaché, United States Embassy, Beijing

"Don't go out there! There's firing in the streets!"[2] Dr. Jonathan Mirsky, the China correspondent for London's *Observer* newspaper, could hear the gunfire even through the massive walls of Beijing's Imperial Palace. Built as a majestic bulwark between the Chinese people and the emperor, these walls seemed as thin as rice paper to Dr. Mirsky when the onslaught began. About midnight on June 3–4, 1989, he had entered the Forbidden City via a side gate and was making his way to Tiananmen Square through a long tunnel. Just then he heard the warning from Robert Delfs of Hong Kong's *Far Eastern Economic Review* and Sandra Burton of *Time* magazine, who had already witnessed the initial shooting.

But Mirsky was determined. A former Dartmouth College associate professor of Chinese, Dr. Mirsky had devoted more than thirty-five years of his professional life to studying China, and he was not going to miss one of the most momentous events in modern Chinese history—no matter how dangerous.

The decision almost cost him his life.

Mirsky moved to a bridge to get a better view of Tiananmen Square—at 100 acres, the largest public square in the world. Towering over the north end was the 120-foot–high Tiananmen (meaning "Heavenly Gate" in Chinese), the entrance to the Forbidden City atop which Mao Zedong had proclaimed the People's Republic of China on October 1, 1949. From his vantage point on one of the Ming Dynasty–era marble bridges that cross over the small moat running in front of the gate, Mirsky could see the twelve-lane Changan ("Eternal Peace") Avenue that cuts through the square.

The main body of the Chinese People's Liberation Army came from Mirsky's right, down West Changan Avenue. First came a single armored personnel carrier followed by a platoon of tanks and companies of infantry. As Amnesty International reported, the armored personnel carrier crushed demonstrators under its treads, "killing and injuring many people."[3]

At the northwest corner, where West Changan Avenue enters the square, the people of Beijing put up what resistance unarmed civilians could muster against the heavily armed soldiers. Mirsky could see sparks on the pavement where the machine gun bullets were bouncing off. A man just to Mirsky's left took a bullet to the chest and crumpled to the ground, a red circle spreading across his white shirt. Tossing Molotov cocktails, some young people on the marble bridges battled squads of paramilitary People's Armed Police that flooded out of the tunnels leading from the Forbidden City. But untrained civilians with homemade gasoline bombs are no match for fully armed and disciplined troops, and the boys on the bridges were beaten to the ground and shot where they lay.

"You m——f—— foreign journalist!" Spotting an inconvenient witness, the People's Armed Police turned on Mirsky and began to beat him with long truncheons. He held onto the marble balustrade of the little bridge, knowing that if he fell, the People's Armed Police would shoot him on the ground. Just when it seemed the end was near, Robert Thomson of London's *Financial Times* and the Italian vice counsel courageously pulled him to safety.

The People's Liberation Army gunned down innocent bystanders; when an ambulance rushed to the scene, the troops murdered the doctors and nurses, too.

Mirsky was temporarily blind and deaf; his arm and jaw were broken; he had lost a tooth, and he had a concussion. But he was alive. Wounded and dying Chinese civilians lay around him.

By 6 AM on the morning of June 4, his hearing and sight returned enough for him to file his story from the Beijing Hotel, two blocks east of the square. Under the headline "Bloodbath in Tiananmen," he wrote:

> The People's Liberation Army smashed its way into the centre of Peking early today, breaching the heart of the people's revolt and starting a bloodbath that left more than 100 dead and hundreds more wounded. Tiananmen Square became a place of horror, where tanks and troops fought with students and workers, where armoured personnel carriers burned and blood lay in pools on the stones.[4]

But Mirsky would witness even more PLA gunfire. At about 10 o'clock that morning, a crowd of anxious Chinese parents seeking word of their children gathered on East Changan Avenue directly in front of the Beijing Hotel.[5] As Mirsky watched in horror, PLA troops opened fire with machine guns, mowing down innocent parents and bystanders. When an ambulance from Beijing's Capital Hospital rushed to the scene, PLA troops murdered the white-gowned doctors and nurses as well.[6] Looking down on the carnage from a Beijing Hotel balcony, Jan Wong of the *Toronto Globe and Mail* counted twenty bodies.[7]

Mirsky's experiences were not unique. For about twenty-four hours beginning on the evening of June 3, the people of Beijing fought the Communist Chinese Army.[8] By the time the PLA armored column reached Mirsky's perch on the north side of Tiananmen Square, it had already gone through at least three major battles with civilians. One of the most significant contests took place at the Muxidi intersection two miles west of the square, beginning at about 9:30 PM. Thousands of civilians had turned buses and heavy trucks into a barricade to block the PLA's route into central Beijing. Armed only with their defiance, the people fended off the army for two hours.

But the end was inevitable. At first the troops were restrained, but at about 11 PM the PLA went on a rampage.[9] PLA armor broke through the flaming bus barricade, and waves of infantry fired on the crowd with their automatic AK-47s. Students from at least four of Beijing's universities died with their school banners flying.[10] A dozen fleeing students were struck in the back.[11] Pierre Hurel of *Paris Match* saw at least thirty people fall before he, too, was hit by a ricocheting bullet.[12]

During a letup in the firing Associated Press reporter John Pomfret jumped on his bicycle and rode eastward through alleyways and small streets parallel to the soldiers' assault down Changan Avenue. Pomfret passed by the next scene of carnage at Fuxingmen overpass, where PLA armor rolled over Chinese civilians, turning them into human paste.[13] At about 11:30 PM Pomfret arrived just west of Tiananmen in time to see the lead armored personnel carriers turn their machine guns on another crowd of civilians. One man next to him was shot in the face, and bodies littered the streets on both sides of the intersection.[14]

The PLA atrocities continued. At about 5 AM the Chinese students holding the southern end of the square negotiated a retreat, but as the marchers crossed West Changan Avenue on their way home, an armored personnel carrier deliberately crashed into the back of the column, crushing eleven students to death.[15] Michael Fathers and Andrew Higgins of London's *Independent* newspaper reported that agonized schoolmates from Beijing's University of Politics and Law packed the bodies of victims in ice and for several days wheeled them from school to school, showing the work of the PLA.[16]

Later on the afternoon of the 4th a platoon of soldiers suddenly broke out of its position east of the square and began shooting at a crowd of civilians. Pomfret and others ran up a narrow alleyway, but the troops fired into the alcoves where they had taken refuge. When the firing stopped, Pomfret could see wounded and dying lying in the streets and alleys.[17]

Other witnesses—who risked their lives to document the atrocities—tell a similar story. At about 3:30 AM on June 4, Colin Nickerson of the *Boston Globe* personally saw at least eight people killed and dozens wounded in a single, five-minute machine gun volley near the Forbidden City.[18] Edward Neilan of the *Washington Times* saw an old man carrying a bamboo bird cage shout at the soldiers, "You're not Chinese, you're animals"; one of the soldiers simply turned his AK-47 assault rifle on the old man.[19]

After rescuing Dr. Mirsky, Robert Thomson of the *Financial Times* returned to the square in time to see the troops begin their final sweep. "When the next round of gunfire began, there was no panic," he reported, but "two youths crumpled within a few yards of me."[20] Between 2 AM and 3 AM a worker from the Capital Iron and Steel Works counted twenty-nine young men and women shot dead as the PLA cleared the square.[21]

On the night of the massacre, a brave woman, the *New York Times*'s Sheryl WuDunn, put on local clothes—so as not to be identified as a Westerner—and went out into the street to get the story:[22]

> Tens of thousands of Chinese troops retook the center of the capital early this morning from pro-democracy demonstrators, killing scores of students and workers and wounding hundreds more as they fired submachine guns at crowds of people who tried to resist.... Most of the dead had been shot, but some had been run over by armored personnel carriers that forced their way through barricades erected by local residents.[23]

Even the Communist Party–owned *Ta Kung Pao* newspaper in Hong Kong condemned the violence. Its June 4 issue appeared with a black border and reported that "scores of people—students and Peking citizens—were shot dead." The editors proclaimed, "Those who have committed this error will come under the judgement of history."[24] Another Communist Party–owned newspaper in Hong Kong, *Wen Wei Po*, reported that Party officials told the PLA troops "not to mind if people are killed."[25] *Wen Wei Po* reported that "the number of residents killed is at least 5,000 while those injured number about 30,000."[26] Before its reporter was yanked off the air, Radio Beijing reported, "Thousands of people, most of them innocent civilians, were killed by fully armed soldiers when they forced their way into the city."[27]

Intense civilian-soldier battles took place all over the city. Just before dawn, veteran war correspondent David Aikman "counted shooting in eight different parts of the City of Beijing, tracer fire ricocheting through the sky."[28] Aikman later reported, "There were sounds of much heavier firing, probably tank shells or very heavy machine guns."[29] According to Aikman, "This was some of the heavier shooting I had ever witnessed as a reporter."[30]

A major struggle surely took place at the Temple of Heaven, a major tourist attraction south and east of the square. The next morning an eyewitness reported that the scene looked like "a civil war had taken place. Everywhere, every intersection, there were burned-out vehicles: jeeps, military vehicles, APCs [armored personnel carriers]."[31] At the Temple of Heaven Hospital the dead ranged from a fifty-nine-year-old factory worker to a twenty-year-old student.[32] Just south of the square, seven people were reported shot early in the evening, the victims of incoming troops.[33]

A four-year-old girl was shot while she held her mother's hand.

Some estimate of unreported battles can be made from the casualty lists at Beijing hospitals. Evidently students had a major confrontation with troops in northern Beijing, as a number of students were hospitalized with severe burns from flamethrowers, a common PLA weapon.[34] Before the PLA ordered officials to cease giving out casualty lists, the hospital reported 95 dead and 125 wounded.[35] And three hospitals in the south of Beijing reported more than a hundred dead.[36] According to British intelligence, General Zhang Wannian's 15th Airborne paratroopers had "mowed down blockading students and citizens."[37]

Beijing's hospitals were unprepared for such carnage. Fuxing Hospital, a small facility closest to the early fighting at Muxidi, was overwhelmed. By morning dozens of bodies were laid out in the parking lot, the oldest sixty-three years of age, the youngest, three.[38] Doctors at Beijing's Capital Hospital reported that a four-year-old girl had been shot while she held her mother's hand.[39]

On one point the PLA and the demonstrators were in agreement: The dead and wounded had to disappear, and quickly. In its account of the massacre, the International League for Human Rights reported:

> The Chinese government has gone to considerable lengths to prevent the true figures from ever emerging. It has prohibited hospitals and mortuaries from disclosing figures of fatalities; troops burnt bodies on the spot in Tiananmen Square, presumably without identification or notification to relatives; and army helicopters were used to airlift bodies and remains to unknown locations.[40]

The wounded wanted to be treated and leave the hospital immediately—a gunshot wound was evidence of counterrevolutionary activity, a serious crime in China. Doctors in Beijing's hospitals colluded with the students to transfer them as soon as possible,[41] and families of the dead collected their relatives as fast as they could for a rapid (and private) cremation.[42] In China today, relatives of dead counterrevolutionaries risk losing all social benefits.[43]

How many people were killed in Beijing on June 3–4, 1989, and in the immediate aftermath? There have been many estimates, but we believe the PLA killed between 4,000 and 6,000 civilians. Early on the morning of June 4, the Chinese Red Cross announced that 2,600 had died, and later that day the Swiss ambassador, who has diplomatic responsibility for the International Red Cross, calculated 2,700 civilian deaths.[44] But as Amnesty International points out, PLA troops continued to open fire on civilians for several days after the June 3–4 massacre,[45] so the number of dead and wounded exceeded the Red Cross's quick count on the morning of the 4th. On June 6, for example, tanks clearing streets for supply trucks opened fire on a group of children, killing two fourteen-year-old boys and a twelve-year-old girl.[46]

Three days after the massacre, NATO intelligence offered an estimate of 7,000 deaths—6,000 civilians and 1,000 soldiers.[47] Some Soviet-bloc estimates were even higher—10,000 killed. A PLA defector in 1996 claimed that a document circulating among military officers had estimated that more than 3,700 people had been killed.[48] But even the PLA doesn't know how many people disappeared into the crematories or died secret, agonizing deaths without medical treatment.

THE VICTIMS

How did it come to this?

The sequence of events that resulted in the Tiananmen massacre began in mid-April 1989 with the death of a disgraced Communist Party leader[49] considered something of a liberal. Beijing university students used the excuse of his funeral to begin pro-democracy demonstrations and marches to Tiananmen Square. They were successful far beyond their expectations. People around the world watched in amazement as a million Chinese people marched peacefully through the streets of Beijing on May 17.

The student leaders proved to be brilliant tacticians. Every time it seemed public enthusiasm was falling off, the students would come up with a new way to draw crowds and spread their message. When momentum from the hunger strike began to fail, art students from the Central Academy of Fine Arts produced the thirty-foot-tall "Goddess of Democracy"—the Statue

of Liberty, with Chinese characteristics.[50] The Goddess immediately succeeded in the students' twin goals of rejuvenating
their movement and offending the old guard PLA leadership.[51]

At the same time the students' message was spreading far
beyond the capital. The government was later forced to admit
that pro-democracy rallies had taken place in 123 Chinese cities
during the spring of 1989.[52] That figure does not even include
Tibet, where severe disturbances occurred in early March.

But what the government feared most was the workers. The
regime could tolerate the demonstrations as long as they stayed
within the circle of university students and intellectuals. A Chinese *Solidarity*, however, would have meant the end of the
regime, and there were signs suggesting that "the Polish disease" (as the Party called it) was spreading.[53]

And indeed the workers were uniting. One charismatic railway worker[54] who joined the students in the square with a small
group of labor supporters was warmly received in Beijing's
industrial centers.[55] When the Communist regime declared martial law on May 20 in response to the protest marchers, his
group formed the Beijing Workers Autonomous Union and
took to Beijing's street corners, giving speeches quoting *The
Communist Manifesto*: "All we proletarians have to lose is our
chains."[56] Similar unions independent of the Communist Party
formed around the country,[57] and an independent labor newspaper was founded as far away as the city of Kunming, 1,200
miles from Beijing.[58]

In the face of this spreading labor discontent, the regime pan-
icked and began to arrest striking workers (but not students)
in late May. But when a thousand angry students showed up at
the Public Security Bureau to show support for the workers,
they were freed.[59] If the workers and students were united, the
Public Security Bureau's police forces weren't going to challenge
them.

THE PARTY

In a Communist regime, the Party is the first line of defense against the people, but in the Tiananmen crisis the Chinese Communist Party (CCP) apparatus fractured. One group decided to support democracy in China while another became accessories to murder. In the early spring of 1989 Communist China's paramount ruler was the old revolutionary soldier Deng Xiaoping, chairman of the Central Military Commission. General Secretary Zhao Ziyang was the titular head of the Party and Deng's chosen successor, but by June 4 he was under house arrest, not to be seen in public again for many years. Zhao had, in effect, sided with the students, which cost him his career. Early on the morning martial law was declared, Zhao met the hunger strikers in Tiananmen Square and expressed sorrow over his inability to control what he knew was to come: "Sorry, we have come too late."[60] After that, he disappeared. Nevertheless, Zhao's staffers enthusiastically supported their boss and the students' cause, at their own cost—one official spent seven years in jail, and another had to flee abroad.

Some Communist officials took to the streets to encourage the demand for democracy. Journalists of the Party propaganda organs such as the Xinhua news agency, the *People's Daily*, and the English-language *China Daily* marched in pro-democracy protests under their own banners.[61] Editors and news reporters of China Central TV issued a statement admitting their errors as "propaganda tools" for the Party and pledged support for the students.[62] And three Party-controlled organizations called for resolution of problems through "democratic and legal means."[63]

In response, other Party officials pressed the old guard leadership for action against the students. Then-Premier Li Peng,[64] a dour, unsmiling, Stalinist-era–educated power engineer, was the public face the Communists put forward at Tiananmen, and as a result he now occupies the commanding heights of the dissidents' pantheon of evil.[65] Li had a famous televised confrontation with student hunger strikers, and he signed the May 20, 1989, martial law decree.

Politburo member and Mayor of Beijing Chen Xitong counts at least as another accessory before and after the fact. Chen briefed the Standing Committee of the Politburo on April 24,[66] and, according to a Tiananmen-era handbill written by anonymous Chinese journalists, he asked the Politburo for "a battle assignment." His report may have led to the hardening of Deng's position and, in turn, instigated an April 26 *People's Daily* editorial that attacked the students.[67] Having partially instigated the violence, Chen signed the first martial law decrees. After the Tiananmen massacre, Chen betrayed his hard-line stance in a speech blaming the entire incident on a conspiracy originated by "some political forces in the West."[68]

It was a Party riven by ideological factionalism and finger-pointing.

THE POLICE

A Communist regime's second line of defense is the internal security organs. They, too, failed at Tiananmen.

In the spring of 1989, internal security was headed by Qiao Shi, a senior Politburo member who had spent his career in intelligence and police work. On June 4 Qiao was not quite in line with the old guard leadership. He had allowed himself to be photographed, smiling, as he visited hunger strikers in the hospital,[69] and by some accounts he abstained in crucial Politburo meetings on the martial law question.[70] He may even have told his police to block the students but not use force against them.[71] The sight of his own Beijing policemen, in uniform, riding around Tiananmen Square saluting the students and giving

them the "V" for "Victory" sign probably influenced his hesitation,[72] as did the sight of Public Security Bureau Academy students who were carrying a banner that proclaimed, "We have
arrived."[73]

The People's Armed Police chiefs had neglected training and
operational readiness.[74] Prior to 1982 the People's Armed Police
had simply been the internal security troops of the PLA, but
after that time it became responsible for border guard duty,
internal security, and even fire fighting. By 1989, of the approximately 600,000 troops the People's Armed Police had nation-
wide, only 30,000 were in Beijing.[75] With its leader (Qiao Shi)
ambivalent, and with strong strains of sympathy for the democracy movement within the police force, the People's Armed
Police was quickly overwhelmed by the student demonstrators.

But after martial law was declared the Communist hard-
liners put the People's Armed Police back under direct PLA control.[76] In essence, they drafted the police into the army. As Dr.
Mirsky discovered, some of the militarized People's Armed
Police units had no compunction against firing on civilians. But
others were absent without leave from the fighting; at one point
a local People's Armed Police detachment assigned to guide
troops unfamiliar with Beijing left the PLA soldiers on their
own in a strange city.[77]

In short, on June 4 the head of the CCP was under arrest,
the Party was split, the secret police chief was on the fence, and
large portions of his police forces were only going through the
motions.

THE ARMY

With rifts in the rest of the regime, the Chinese People's Liberation Army became both the originator and the instrument of
repression. Three PLA politicians in civilian clothes—Deng

Xiaoping, Yang Shangkun, and Wang Zhen (all now deceased)—made the decision to destroy China's democracy movement. Six active-duty uniformed PLA generals—Chi Haotian, Xu Huizi, Yang Baibing, Li Jijun, Kui Fulin, and Xiong Guangkai—were in Beijing and were instrumental in carrying it out. A seventh, Zhang Wannian, was not in the city for the June 3–4 massacre—he was stationed in southern China to quash any resistance in that region—but his paratroopers conducted some of the most gruesome assaults on the Chinese people. Everyone of any consequence in the massacre was a long-serving military officer.

Deng Xiaoping was chairman of the CCP's Central Military Commission. But that was not his only connection to the PLA. As Deng's daughter, Deng Maomao, points out in her biography of her father, by the time the PRC was proclaimed on October 1, 1949, Deng had spent more than

In the eyes of the hardline leaders, the People's Liberation Army had to make China safe *from* democracy.

twenty years in the PLA, nearly all his adult life to that point.[78] Had the Communists used normal military rankings, Deng would have been a four-star general by 1949. Attacked as a "rightist" by Mao and his Red Guards, Deng and his family suffered tremendously during the Great Proletarian Cultural Revolution (1966–1976). He was brought back to power twice by his old army comrades who felt the Party needed his firm hand. From 1977 to 1990[79] Deng had some military rank, either as PLA chief of staff or as chairman of the Central Military Commission.

Another key decision-maker was General Yang Shangkun, vice chairman of the Central Military Commission and president of the PRC, a largely ceremonial post. General Yang had been in the PLA since the early 1930s.[80] During much of his career Yang was the commander of the elite unit of bodyguards assigned to protect the Communist Party leadership.

General Wang Zhen,[81] vice president of the PRC, was an experienced battlefield commander and former deputy chief of staff of the PLA. General Wang was a vicious, hard-line Communist (former head of the Central Party School) known for his ideologically extreme tendencies.[82] In 1949 he led PLA troops to suppress Muslim separatists in China's far western province of Xinjiang (Chinese Turkestan).[83] Under Wang the PLA crushed Islamic life in Xinjiang—mosques and schools were closed, religious leaders and teachers were imprisoned or exiled.[84]

Those in uniform in 1989 were hard-liners as well. General Chi Haotian, then PLA chief of staff, was in operational command of the troops at Tiananmen,[85] while General Xu Huizi, then the PLA's deputy chief of staff, was in tactical command.[86] General Yang Baibing, half-brother to General Yang Shangkun, was chief of the PLA's General Political Department, which in essence is the Communist Party's control apparatus over the PLA, since the PLA's political commissars report to it. General Li Jijun was Deng Xiaoping's military aide, while General Kui Fulin was the PLA's operations director and thus directly responsible for planning the massacre. Finally, General Xiong Guangkai was the chief of military intelligence and ran provocation operations against the students.

(The Clinton-Gore administration later welcomed to Washington five of these six generals—all but Yang Baibing—as well as Zhang Wannian, whose paratroopers mowed down civilians at Tiananmen. In most cases the PLA generals, responsible for the deaths of *thousands* of Chinese, received high honor greetings at the Pentagon, and some, including Zhang and Chi, were granted White House meetings with President Clinton.

Certain members of the Deng and Wang families also made controversial appearances on our shores.)

Given the support for democracy across Chinese geographic, generational, and social lines, the Beijing regime was in serious trouble in the spring of 1989. General Yang Shangkun told the PLA leadership that the demonstrators' "purpose was to overthrow the Communist Party and the current government."[87] According to Yang, "If we retreated, everything would collapse."[88] In the eyes of the hard-line leaders, Yang and his PLA comrades had to make China safe *from* democracy.

AN ARMY THAT KILLS ITS OWN

During what became a two-month PLA military operation, the United States and other NATO military attachés had, in effect, a front-row seat.[89] As British military analyst Tai Ming Cheung points out:

> Throughout the crisis, the U.S. and its NATO allies pooled resources in keeping a constant watch on the PLA's activities, in particular the mobilization and deployment of military units into Beijing. Maintaining the rule that only verified information—that which attachés saw firsthand—was to be reported, the Western diplomatic and intelligence community felt that it pieced together a solid and detailed picture of military deployments in and around Beijing.[90]

The PLA leadership became so exasperated at the attachés' excellent reporting that three days after the Tiananmen massacre an armored personnel carrier stopped in the middle of Changan Avenue and specifically targeted the living quarters of American diplomats with machine gun fire.[91] In one apartment the parents were not at home at the time, but their small children were in the living room in the care of their Chinese *Ai-yi* (nanny). Although it is widely assumed that Chinese house servants in the employ of foreigners also work for the Chinese secret police, the *Ai-yi* threw herself on top of the children as machine gun bullets broke glass two feet over her head. Fortunately, neither the children nor their protector was injured.

After eighteen bullet holes were counted in the American officer's quarters,[92] U.S. Ambassador James Lilley told the Chinese vice foreign minister that if any Americans were harmed by the PLA, the United States would break relations. There were no more incidents.[93]

It seems likely that the first serious military order against the dissidents was given on April 25.[94] With Zhao on an official visit to North Korea, General Yang Shangkun went to visit Deng.[95] The two old soldiers discussed the situation, and Deng reportedly indicated a lack of concern over shedding blood or any possible international reaction.[96]

Deng and Yang initially decided to bring the 38th Group Army into play. This Group Army, whose mission is to defend the capital, was fully equipped and at full strength, with two infantry divisions and an armored division. Deng's order on April 25 brought ten thousand troops on the march from the 38th's headquarters seventy-five miles south of Beijing.[97]

But there was still the question of loyalty. The people of Beijing thought of the 38th as "our army" because of its proximity to the capital and because many Beijingers had done their military service with it.[98] In fact, during the chaos of June 3–4 there were street rumors (false, it turned out) that elements of the 38th had refused orders or even turned their guns on other PLA units in defense of the demonstrators.[99]

Between April 25, when Deng gave his first serious military orders, and May 20, when martial law was imposed, the PLA leadership faced a deteriorating situation. While its opposition was organizing and strengthening, the PLA leadership was looking at a fractured Communist Party, a collapsing internal security apparatus, and a potentially disloyal elite military unit, the 38th Army.

In response, the three old guard leaders—Deng, Yang Shang-kun, and Wang Zhen—ordered every military region to send some forces to Beijing, no matter how remote the region or how unsuited its forces were to the task at hand.

General Wang Zhen became the voice of the old guard calling for the strongest measures. RAND Corporation China scholar Michael Swaine describes General Wang as having a "clear penchant for repression" and suggests he "was a key proponent of the Tiananmen crackdown."[100] Brigadier General Michael T. Byrnes, formerly head of the United States Defense Liaison in Beijing, also points to Wang as one of "the most likely candidates to use the PLA," noting his prominent protocol position on May 20, when martial law was announced, and on June 9, when Deng reported on the massacre.[101] One unconfirmed report even had General Wang on West Changan Avenue giving the soldiers "shoot-to-kill" orders during the massacre.[102]

Time magazine's David Aikman judges that, as operational commander, Chief of Staff Chi Haotian "bears major responsibility for the violence unleashed upon Beijing's citizenry."[103] Chi set up his headquarters at the PLA Military Museum on West Changan Avenue just beyond the Muxidi intersection.[104] He could deploy any number of units to Beijing, but he was not without his own circle of old military comrades to draw on; during his career he had been the political commissar of the 27th Group Army, which would feature prominently in the massacre.

General Chi took a month to deploy his troops to the Beijing area by air, rail, and truck. The United States Defense Liaison Office in Hong Kong estimates that by the night of the massacre, at least fourteen of the PLA's twenty-four Group Armies and two airborne brigades were represented in the Beijing area, totaling fifty thousand men. General Byrnes, who was in Beijing on June 4, has a higher estimate—sixteen to eighteen Group Armies represented.[105] Although most Group Armies, particularly those headquartered far from Beijing, sent only one division, all Group Army commanders and political commissars and their deputies were present with their troops.[106]

> What was at stake was the survival of Communism in China—and the hardline leadership was willing to commit any crime to preserve it.

General Chi had assembled a mighty force, but the question was, would they fire on the Chinese people? The PLA leadership had to maintain unity and discipline over three groups: the troops, the officer corps, and the top commanders.

General Yang Baibing, chief of the PLA's General Political Department, drew the first assignment. He and his political commissars had to prevent the enlisted troops coming in from outlying areas from becoming infected with the democracy disease. In practice, this meant keeping the People's Army away from the people. This directly contradicted Mao's most basic military teaching, which called for unity between the people and the army. According to Mao, "The People's Liberation Army men are like fish in the sea of the people."[107] Yang had to prove that PLA fish could live out of water. His only hope was to separate his troops from the people and expose them to intense propaganda.

While the young troops were undergoing intensive ideological training on the outskirts of town from Yang's political commissars, PLA officers held their own meetings. After graduation from military academies, PLA officers are scattered all around China, an area larger than the United States (including Alaska and Hawaii). The Tiananmen operation was the first time many of them had seen each other since graduation, so the various academy classes took over local restaurants to socialize. Reestablishing old connections in the officer ranks created a new military solidarity for this mission.[108]

General Yang Shangkun made the case for military intervention before the top PLA brass on May 24. According to Yang, "At present, Beijing is in a chaotic condition... [and] is now out of control."[109]

But in fact, the opposite conditions prevailed. On more than one day, a million citizens of Beijing had held joyous celebrations of liberty without a single casualty or even a broken window. China in the spring of 1989 probably had its lowest crime rate ever: Even the thieves declared a moratorium on their activities and joined the students in Beijing and Xian.[110]

What was at stake here was the survival of Communism in China—and the PLA leadership was willing to commit any crime to preserve it.

The armies were assembled, and solidarity was achieved at all levels—troops, officers, and generals. All that was needed was the order.

General Xu Huizi, the deputy chief of staff and tactical commander for the Tiananmen operation, called his senior officers together at 4 PM on June 3 for their final orders.[111] General Xu directed three waves of mechanized infantry to jump off at two-hour intervals, backed by a fourth wave of heavy armor.[112] The soldiers would originate from all points on the compass and meet up at Tiananmen Square. Group Army commanders were pitted against each other in a contest to see who could reach the square first.[113]

All of the forces met strong resistance from Beijing residents. In some cases crowds of thousands brought military movement to a halt.

But General Xu broke the back of the civilian resistance with a full-scale mechanized infantry assault. Led by armored personnel carriers firing .50 caliber machine guns and backed by heavy armor, PLA troops rolled down West Changan Avenue, smashing any resistance in their path. In 1956 the people of Budapest had fought Soviet armor with light infantry weapons donated by the Hungarian army, but the people of Beijing had purposely turned in any PLA weapons that had come into their hands.[114] Their cause was just, and they felt nonviolence would prevail. But by 3 AM on June 4, the PLA was in control of the square.

Xu used Beijing's own 38th Group Army, supported by the 27th. The people of Beijing didn't want to believe their own 38th would fire on them, which explains the fantastic stories that went around saying that soldiers from the 38th had turned their guns on the 27th.[115] Only when British intelligence could examine the main battle tanks of the armored column did the 38th's betrayal become apparent: At the time only the 38th Group Army was equipped with China's newest tank, Type 69, and the tanks in Tiananmen Square were Type 69s.[116]

Yet the people of Beijing had been partially right about the 38th. On June 3 the 38th's commander was under arrest for refusing to support martial law, and he was later court-martialed and imprisoned.[117] Nevertheless, his deputy forced the soldiers of the 38th to obey PLA orders, and for his loyalty to the Communist regime he succeeded to command of the 38th.[118]

By nearly all outside accounts,[119] the 27th Group Army was responsible for most of the killing. Some observers speculated that its ferocity was generated by its experience in the 1979 war with Vietnam.[120] But if the 27th was the hammer that fell on Beijing, it was because General Chi Haotian felt his old unit was the most reliable.

CRUSHING THE OPPOSITION

Speaking before a friendly audience at the U.S. National Defense University in late 1996, General Chi, by then China's defense minister, declared not only that no one had died in Tiananmen Square itself, but also that there had been only some minor "pushing and shoving" around the periphery.[121]

Having crushed the dissidents and secured Communist power in China, the PRC leadership is free to offer such blatant lies. Only by looking at the *real* history can we understand the extraordinary brutality of the Beijing regime. If the Communist Chinese can gun down thousands of innocent civilians in order to ensure the regime's survival, it is only a matter of time before they turn their guns on the rest of the world to satisfy their territorial ambitions.

CHAPTER 3
A PARTY ARMY

The truth, of course, is that the uprising was the greatest show of democratic force in over forty years of Communist rule in China.[1]

—Liu Binyan

The PRC Embassy in Washington, D.C., is a 1950s-era pile on the edge of Rock Creek Park. The embassy was once a hotel, and still provides living quarters as well as offices for embassy personnel, which makes it much easier for the Chinese secret police to keep an eye on them.

The morning after the Tiananmen Square massacre, crowds of demonstrators began to gather in the little triangular park across from the embassy. Looking up at the front of the building, the authors could see that, in solidarity with the demonstrators, Chinese diplomats had tied the ugly beige window curtains together, making them "V"s for "Victory."

The diplomats paid for their symbolic bravery. Late one night in early July, a column of buses lined up at the embassy's front door to return these Chinese patriots to their superiors in Beijing.[2]

WHERE COMMUNISM LIVES ON

On January 1, 1989—the year Chinese patriots ignited popular rebellion against Communism[3]—nearly two billion people lived under Communist rule.[4] Five months after Chinese hopes were crushed at Tiananmen, the Berlin Wall fell. By the end of 1991 the Soviet Union was extinct and Communist regimes from East Berlin to Ethiopia to the Pacific Ocean had been overthrown. Five hundred million people no longer lived under Communism.

The Chinese Communist Party's internal security apparatus can now quash any Tiananmen-like protest before it can even get started.

But in China, where the dissidence began, the rebellion failed.[5]

Two questions immediately come to mind: First, how did the Chinese Communist Party (CCP) leadership survive from 1989 to 1991 when other Communist regimes did not? Second, how does the CCP maintain its power now, in the face of the unprecedented contact between the Chinese people and visitors from democratic countries?

At the conclusion of his December 1996 visit to the U.S. National Defense University, General Chi Haotian, now the PRC's defense minister, assured American military officers that Tiananmen would never be repeated in China. Some thought he meant the PLA would never again fire on the Chinese people. In fact, he meant something more ominous: The CCP's internal security apparatus can now quash any Tiananmen-like protest before it can even get started.

PUNISHMENTS AND REWARDS IN THE WAKE OF TIANANMEN

In the disarray that immediately followed June 3–4, 1989, the Beijing leadership turned to the army. On June 9 Deng Xiaoping addressed the Martial Law Command and commended the PLA as the Party's "Great Wall of Steel." Beijing remained under martial law until the spring of 1990. During that time, the PLA not only patrolled the streets but also controlled the news. PLA officers were installed in every media outlet, including the *People's Daily*, the CCP newspaper.[6]

Within a week of the massacre, "Most Wanted" posters went up around the country; government prosecutors were told not to worry about legal niceties in detaining opponents of the regime.[7] About half of the highest ranking student leaders managed to escape in the summer of 1989, but by 1992–1993 the Public Security Bureau—the regular police—had been reorganized and had arrested many dissidents.[8] Reporting seven months after the Tiananmen atrocities, the U.S. Department of State declared:

> The Beijing massacre was followed by a drastic, country-wide crackdown on participants, supporters, and sympathizers. Thousands were arrested and about a score are known to have been executed, following trials which fell far short of international standards, for alleged crimes committed during the unrest.[9]

As late as nine years after the massacre, three thousand people were still in jail for offenses related to Tiananmen. One young boy was serving a fifteen-year prison term for buying matches.[10]

The Party was suitably grateful to its saviors. PLA generals who had been willing to shoot unarmed civilians were promoted. After proving his loyalty to the Party at Tiananmen, Chief of Staff Chi Haotian became a state councilor, defense minister, and a member of the Central Military Commission.[11] General Yang Shangkun moved up from executive vice chairman to first vice chairman of the Central Military Commission, and his half-brother Yang Baibing became its secretary general. Within two years 70 percent of the leadership in the military regions changed—mostly by promotion.[12] New Party Secretary and Central Military Commission Chairman Jiang Zemin began to hand out promotions to his generals at an unprecedented rate; nineteen officers became full generals in June 1994 alone.[13]

> China remains a nation held captive by the Chinese Communist Party and its military arm, the People's Liberation Army.

Rewards are still being handed out. In May 1999 a general associated with the Tiananmen massacre was named political commissar of the PLA troops in Hong Kong.[14]

The PLA as an institution was also rewarded. PLA membership in the Communist Party Central Committee went up from 18 percent to 22 percent,[15] and Deng issued secret orders to allow ten generals to sit in on Politburo meetings.[16] This gave the PLA broader political participation at the center than it had had in decades. In addition, the PRC reversed years of declining defense budgets with a 15.2 percent increase in 1990.[17] The PLA air force received its first modern third-generation jet fighters, Russian Su-27s, in 1992. And money began to flow into weapons production; according to PLA figures, missile production rose 53 percent from 1992 to 1993.[18] It was all part of an unprecedented military buildup.

Moreover, the PLA became a major player in state-controlled business interests. According to noted China authority Willy Wo-lap Lam, at a "landmark CMC [Central Military Commission] meeting in January 1993 Jiang gave the green light for the PLA to pursue business on a large scale."[19] Within six months PLA businesses had absorbed a billion dollars' worth of foreign investment.[20] Rampant corruption allowed the senior officer corps to siphon off hundreds of millions of dollars.[21]

While the PLA held down dissent, Beijing turned its attention to reestablishing the Communist Party apparatus, which was apparently losing ground among the people to religion. In early 1991 hard-line General Wang Zhen revealed that the people were seeking answers in Christianity, not Marxism. In one county in central China 813 people became Catholics while only 270 chose to join the Party.[22]

From 1989 to 1991 Deng had turned leftists loose on the Party. In the first half of 1990, under security boss Qiao Shi's direction,[23] the CCP required all Party members to reregister in order to examine their loyalty and ideological purity. As a result, more than 100,000 names were dropped from the rolls.[24] Government, factories, and universities had new Party organizations imposed on them,[25] and the CCP geared up its propaganda machinery.[26]

HUMAN RIGHTS AND DEMOCRACY IN CHINA TODAY

China remains a nation held captive by the Chinese Communist Party and its military arm, the People's Liberation Army. In 1996 the U.S. State Department reported:

> The [Chinese] Government continued to commit widespread and well-documented human rights abuses, in violation of internationally accepted norms.... All public dissent against the party and government was effectively silenced by intimidation, exile, the imposition of prison terms, administrative detention, or house arrest. No dissidents were known to be active at year's end.[27]

In its 1998 human rights report, the State Department declared, "The [Chinese] Government's human rights record deteriorated sharply beginning in the final months of the year with a crackdown against organized political dissent."[28] No one knows how many Tiananmen-era people are still languishing in prison camps, the State Department reported, because "the Government still has not provided a comprehensive, credible accounting of those missing or detained."[29]

> The People's Liberation Army is willing to kill Chinese people in order to maintain a Communist regime in power.

The PLA has even gone so far as to harvest executed prisoners' vital organs for sale to the highest bidder, as ABC's *Prime-Time Live* revealed on October 15, 1997. At great personal risk, ABC's team of Brian Ross and Rhonda Schwartz entered a restricted military zone in southern China to show that the PLA is selling body parts to those wealthy enough to afford them. According to ABC, the Chinese military has sold "perhaps thousands" of kidneys since the late 1980s.[30]

SERVING THE PARTY

What kind of military establishment would allow itself to be used against its own countrymen? A Party army. The PLA is directed by the national constitution to be loyal to the Communist Party of China.[31] PLA dedication to the Party comes before all else. "The Chinese Communist Party has absolute authority over the Army," PRC President Jiang Zemin told the PLA's newspaper in 1997.[32]

This relationship is natural given that the CCP came to power at the point of a PLA bayonet—and has remained in power thanks to it. For example, during the chaos of the 1966–1976 Cultural Revolution that split the Party, Mao used the PLA to save the CCP from itself and restore order in the country. The PLA was also called out to put down a Muslim rebellion in South China, using artillery to kill five thousand people.[33] Today, the PLA keeps the Party in power. The CCP has never held public elections, a deplorable but understandable decision given that Democratic Party leader Martin Lee regularly trounces the Communists in Hong Kong elections.

It is, therefore, not surprising that after the Berlin Wall fell in 1989, the CCP survived in power when other Communist parties did not: *The People's Liberation Army is willing to kill Chinese people in order to maintain a Communist regime in power.*

Would the PLA crush another pro-democracy movement in China? Without a doubt. In the words of Professor David Shambaugh, former editor of *China Quarterly*, "The [Chinese] army is not going to let the [Communist] Party fall from power as was the case in the Soviet Union."[34] The U.S. Army agrees. In its "China Battlefield Development Plan," a Defense Intelligence Reference Document, the army states, "The role of China's armed forces is to defend China's sovereignty and territorial integrity... and to preserve the communist system."[35] The U.S. Army believes the "primary mission" of the PLA is "support of the CCP against internal unrest."[36]

THE PEOPLE'S ARMED POLICE

With regard to internal security, the PLA has two goals: (1) make certain China does not become a democratic country, and (2) place the blame for repression on someone else. The unit designated for these two missions is the People's Armed Police, the same group that tried to murder reporter Jonathan Mirsky on the night of the Tiananmen Square massacre.

For a long time People's Armed Police troops were the orphans of the Chinese military system.[37] Command of its troops shifted back and forth between the PLA and the Public Security Bureau; at one time command even shifted to Communist China's rocket forces. And its troops were last in line to receive new equipment and financial support from the central government.

But all that changed after Tiananmen. In order to prevent democracy from overthrowing Communism in China, the CCP revamped and expanded the People's Armed Police. Because of its importance to the regime's survival, the People's Armed Police went from last on Beijing's priority list to near the front:

Leadership: The new head of the People's Armed Police is a tough PLA general with extensive battle experience fighting American troops in Korea.[38] Other PLA officers have also been transferred to the People's Armed Police in order to beef up its experience and discipline.

Command: In 1995 the People's Armed Police was placed under the direct command of the Party's Central Military Commission.[39]

Personnel: The People's Armed Police has nearly doubled in size since 1989, from approximately 500,000 troops to close to a million.[40] After Tiananmen, entire divisions of PLA troops simply changed uniforms.[41]

Budget: The budget of the People's Armed Police is guarded closely in Beijing, but China watchers generally believe that it has increased substantially, if only to pay for the new personnel and equipment.[42] The People's Armed Police has also recently participated in the PLA's commercial operations.

Equipment: The People's Armed Police has received new tanks, artillery, armored personnel carriers, armored cars, helicopters, tear gas, rubber bullets, riot shields, electric prods, trucks, and satellite communications gear, among other improvements.[43]

Tactics and Training: The Chinese press is full of breathless accounts of the new People's Armed Police night training: "Sharpshooters' laser beams accurately mark the criminal's vital midsection, and a crisp rifle report of 'Ping, ping, ping...' rings out."[44] There is also some discussion of People's Armed Police "fist" or "rapid reaction" forces in training.[45]

The American defense secretary's own China specialist has noted, "By increasing the size of the People's Armed Police, the leadership in Beijing implicitly acknowledges that internal unrest is a greater threat to the regime's survival and Chinese economic modernization than is foreign invasion."[46] The People's Armed Police has recently seen action in quashing such internal unrest, handling what the Party refers to euphemistically as "unexpected incidents"[47]—putting down Muslim demonstrators in western China,[48] killing farmers who protest high taxes and corruption in southern China,[49] and suppressing religious groups anywhere in China and Chinese-occupied Tibet.[50]

RAPID REACTION

The PLA leaves nothing to chance. Army soldiers in police uniforms—the People's Armed Police—may be the first line of defense, but in the end the PLA knows it may be called on again. Since the Tiananmen massacre the PLA has set up its own "rapid reaction" forces modeled on General Zhang Wannian's 15th Airborne paratroopers that were the shock troops at Tiananmen.[51] These highly trained, mobile, elite troops, outfitted with top-of-the-line equipment, are responsible for getting to the scene of a disturbance quickly to snuff it out, with all the brute force necessary.

HELD AT GUNPOINT

Even more than twenty years after his death, Mao Zedong's dictum that "Political power grows out of the barrel of a gun"[52] remains the central political principle of modern China, and the PLA is its enforcer. To retain the CCP's power, the PLA has turned its guns against the Chinese people in the past, and it may do so again.

Worse still, the PLA has shown that it has its sights set beyond China's own borders. Its history of aggression could spell disaster in the coming century.

CHAPTER 4
A HISTORY OF AGGRESSION

Every Communist must grasp the truth: Political power grows out of the barrel of a gun.

—Mao Zedong

I n the Tibetan capital of Lhasa, a stone obelisk erected in AD 823 features a treaty between China and Tibet:

Tibet and China shall keep the country and frontiers of which they are now in possession. All to the east is the country of Great China; and all to the west is, without question, the country of Great Tibet. Henceforth on neither side shall there be waging war nor seizing of territory. There shall be no sudden alarms and the word "enemy" shall not be spoken.... This solemn agreement has established a great epoch when Tibetans shall be happy in the land of Tibet and the Chinese in the land of China.

The treaty has been ignored by the Chinese Communist Party (CCP). Tibet is a nation under Chinese military occupation.[1]

In the fall of 1950 the Chinese People's Liberation Army invaded and occupied Tibet. The PLA remains there today. The country has been dismembered, with half its territory assigned to Chinese provinces. Central Tibet is an armed camp that the PLA and its paramilitary police hold only by sheer weight of numbers. At least a million Tibetans—one out of every five or six—have lost their lives to Communist Chinese terror.

Today the Chinese Communists hold on to Tibet by brute force—nothing else.

The Tibet of 1950 saw no communal violence or civil war between rival claimants for power, no streams of pathetic refugees throwing themselves on the mercy of neighboring countries. Contrary to fifty years of Communist propaganda, Tibet was not in need of "liberation."[2] The PLA launched an unprovoked attack on defenseless people.

The PLA began with border raids, probing Tibet's defenses, followed by a full-scale invasion with eighty thousand battle-hardened troops.[3] As His Holiness the Fourteenth Dalai Lama noted in a personal letter to one of the authors:

> Tibetans could not possibly resist such an onslaught. The Tibetan army, totaling 8,000 officers and soldiers, was poorly trained and equipped. It was no match for the PLA, who quickly encircled the Tibetan army and smashed the entire defense line.

The Dalai Lama was fifteen years old when his country was conquered. For a time in the early 1950s he was more or less a prisoner in Beijing while the Communists secured Tibet. In 1956, led by a number of women,[4] the people of Tibet began to rebel,[5] but the PLA brought in tens of thousands of reinforcements.[6] This, in turn, led to another cycle of violence and, finally, open revolt in 1959. But with an inexhaustible supply

THE DALAI LAMA

Dear Mr. Triplett,

I would like to commend you for the valuable work you have been doing to highlight the situation in China and in Tibet, and to support the aspirations of the Chinese and Tibetan people.

I would like to share with you a few of my thoughts about the Chinese invasion of Tibet, and the current state of our relations.

At the end of 1949, the newly established People's Republic of China announced in Peking that they would "liberate" Tibet from the hands of the Western imperialist powers. At the time Tibet was a fully independent nation with only half a dozen Westerns in the entire country. After taking over Tibet's northern and eastern parts, in October 1950, 80,000 soldiers of the People's Liberation Army crossed the Drichu River east of Chamdo into Central Tibet.

Tibetans could not possibly resist such an onslaught. The Tibetan army, totaling 8,000 officers and soldiers, was poorly trained and equipped. It was no match for the PLA, who quickly encircled the Tibetan army and smashed the entire defense line. After defeating the Tibetan army, Tibet was forced to negotiate with China and signed the Seventeen Point Agreement under duress. The situation rapidly deteriorated after that, leading to the tragic events in 1959 at which time I was forced to leave my country.

Since then, my country has remained under the occupation of the People's Republic of China. I have made countless efforts to peacefully resolve this situation through negotiations with the Chinese Government, but have thus far, not succeeded.

Tibet is an important buffer and I remain committed to the proposals I have made to establish Tibet as a zone of ahimsa. This is critical to establishing peace in Tibet, and in the entire region.

I hope this information is helpful to your work.

With best wishes,

The Dalai Lama

Mr. William Charryl Triplett, II

This personal letter from the Dalai Lama to the one of the authors describes the PRC's bloody invasion of Tibet in 1950.

of PLA soldiers at Beijing's disposal, the issue was never in doubt. At least 87,000 Tibetan freedom fighters lost their lives defending their homes and country over an eighteen-month period, and the Dalai Lama was forced to lead his people into exile in India.[7]

After nearly a half-century of PLA rule, modern Tibet reels under the influence of its Communist Chinese oppressors. Strategically placed around Lhasa's open square are surveillance cameras and special microphones designed to eavesdrop on the population. The Communist Chinese secret police are easy to spot among the throngs of people—they're the street thugs in leather jackets. Similarly, overwhelming numbers of PLA soldiers in their dirty green uniforms are a reminder that Tibet is an armed camp. Chinese police vehicles travel the roads outside of Lhasa, lights flashing, sirens blaring, forcing the old buses filled with Tibetans to the side of the road. Electric power and telephone lines pass over the Tibetans' adobe compounds to service the cinder-block buildings housing the Chinese.[8]

Communist China's continued human rights violations in Tibet—executions, imprisonment, torture, rape, forced abortions, and cultural annihilation—are all well documented.[9] Assaults on Buddhist nuns with electric cattle prods seem to be a particular amusement for the Chinese People's Armed Police.[10] Of more than six thousand religious sites existing in 1950, only a handful remain.[11] Thousand-year-old libraries were put to the torch by Chinese mobs during the Cultural Revolution, an entire cultural history of a people, lost.

Unfortunately, the executive branch of the American government does not recognize Tibetan independence. Congress, however, does—and Congress is right. In 1987 Congress passed the first modern policy statement on Tibet, accurately declaring that the PLA had "invaded and occupied" Tibet.

As early as 1922 the Chinese Communists declared their intention to "liberate" Tibet.[12] Today they hold on to Tibet by brute force—nothing else. Without the overwhelming presence of the PLA, Chinese rule would end tomorrow. Hundreds of thousands of active-duty PLA soldiers backed up by armor and highly trained paramilitary troops form the backbone of the occupation forces. In addition, demobilized soldiers and People's Armed Police troops have been transferred to Tibet as civilian officials in Party and government organizations.[13]

Why did the CCP send its military arm, the PLA, to inflict such punishment on an innocent neighbor? Probably a combination of reasons—some Communist revolutionary fervor, some Chinese chauvinism—but most likely because control of Tibet gives the Chinese the high ground from which they can threaten all of South Asia. Historic Tibet bordered on a wide sweep of territory from Burma to Pakistan. From Tibet's heights, the PLA can mount direct military attacks and armed subversion almost at will.

> **The People's Liberation Army has engaged in armed aggression or subversion against almost all of China's neighbors.**

The PLA's seizure of Tibet was not an isolated phenomenon. In the fifty years since the PRC was established, the PLA has engaged in armed aggression or subversion against almost every one of China's neighbors—Tibet, South Korea, India, the Soviet Union, Vietnam, Laos, Cambodia, Thailand, Burma, the Philippines, Indonesia, Taiwan, Malaysia, and Singapore. Victims count in the millions; pain and destruction have been widespread.

It is a frightening record of insatiable territorial aggression—a record the Clinton administration has ignored as it has sought to form a "strategic partnership" with the PRC.

KOREA, 1950

Captain Joseph Darrigo, an American adviser to the South Korean army, awoke to the sound of cannon fire. It was 3:30 AM in Seoul on Sunday, June 25, 1950, and he was the only American officer at the 38th Parallel. After an hour-long artillery barrage, waves of battle-hardened North Korean infantry rose out of long-prepared positions and swept over the southern defenders, led by 150 T-34 Soviet-built tanks.[14]

What Darrigo and the South Korean defenders did not know was that the lead divisions of the North Korean attack had received their combat experience as members of the Chinese People's Liberation Army.

After World War II, Korea had been divided, with the Soviets controlling the North, and the Americans, the South.[15] North Korea, which the Soviets converted into a militarized, Communist state, in due course became the rear support area for the PLA's conquest of Manchuria.[16] Entire North Korean army divisions—totaling some 100,000 soldiers—fought for years in PLA uniforms against the Chinese Nationalist forces.[17] Some historians believe that these North Korean soldiers were the key to the PLA's victory in Manchuria[18]—a Communist victory that put the Nationalists on the road to ultimate defeat.

In the summer of 1949, when a Communist victory in China seemed assured, North Korea's Communist dictator took back his troops. Estimates vary, but at least 50,000 to 70,000 fully equipped ex-PLA combat troops returned to North Korea, forming the nucleus of the 135,000-man army that invaded the South without warning in June 1950.[19]

In contrast to the formidable North Koreans, the South Korean forces were poorly trained, inexperienced, and ill-equipped. Although South Korea supposedly had 100,000 troops, in reality most units were not at full strength, a problem compounded by the fact that many soldiers were in town on pass when the North Koreans invaded. The South Korean troops had no armor, no anti-tank weapons, no air cover, and no heavy artillery support because the American State Department had vetoed weapons sales to South Korea, believing such sales would be provocative[20]—a position the State Department now takes with regard to some arms sales to Taiwan.

Facing a surprise armored attack supported by division after division of heavily armed, disciplined, and experienced infantry, Darrigo's single regiment of South Koreans didn't stand a chance. The underequipped South Korean army—known as the "ROKs," for Republic of Korea—retreated.

Within ten days of the outbreak of war, American ground troops were engaged in Korea, fighting side by side with the ROKs. But by the beginning of August, there was nowhere to retreat. With their backs to the sea, the ROKs and the Americans fought off wave after wave of North Korean attacks on the South Korean port of Pusan. Finally, on September 15, seventy thousand United Nations (UN) soldiers and marines under the command of General Douglas MacArthur landed behind enemy lines at Inchon, the port of Seoul. Within days, Seoul was freed of Communist rule, the North Korean army barely existed as a unified fighting force, and the UN was marching north.

But there was no panic in Beijing as it watched the Free World defeat its North Korean ally, because, as we now know, the Communist Chinese leadership had begun to prepare its next move in the first week of July, moving troops from the southern part of China to the north. By mid-October tens of thousands—soon to be hundreds of thousands—of PLA soldiers were crossing into North Korea. For the next eight months UN forces would endure five major Chinese military campaigns.[21] Seoul would change hands a total of four times before the PLA was finally thrown out.[22]

For almost fifty years there has been intense debate about the PLA's intervention in Korea. For a long time Western academic circles held that it was somehow "our fault"—that is, the Free World in some way provoked the Chinese.[23] But with the demise of the Soviet Union and the opening of the Russian state archives, it's now fairly clear that, from the beginning, Beijing was deeply involved in the attempted conquest of South Korea. Mao and his comrades were perfectly happy to have the North Korean Communists take the lead, but they were equally prepared to intervene when necessary. In short, the anti-democratic rhetoric coming out of China in those days was not empty propaganda; Mao, Stalin, and the Korean dictator truly intended to establish Communism all over Asia—by force, if necessary.[24]

The PLA itself suffered terrible losses at the hands of the American soldiers and airmen. According to the UN, there were 900,000 Chinese casualties, including 300,000 killed, but even those numbers may be low.[25] The Communist leaders experienced the devastation firsthand: a son of Mao Zedong was killed by the U.S. Air Force.[26] But such losses are the price the Communist Chinese are willing to pay to pursue their expansionist agenda.

What effect did the PLA's intervention have on the Korean peninsula?

- At least one million Korean civilians lost their lives.
- More than fifty thousand American soldiers, sailors, airmen, and marines were killed.
- A vicious Communist state, possibly nuclear armed, now holds half of Korea.

- North Korea's missile, chemical warfare, and germ warfare programs threaten Japan, India, and American friends and allies in the Middle East.
- For almost fifty years, the United States has been forced to keep forty thousand soldiers on alert against another invasion of the South.
- The Korean peninsula remains near the top of any expert's list of potential flashpoints.

INDIA, 1962

The Indians fought to the last man. When the final round of ammunition was gone, they met the invaders with the bayonet. There were no deserters.[27]

But, like Captain Darrigo's South Koreans, the Indians had no chance to defend themselves against the Chinese. The PLA attacked by surprise and from three sides in a well-coordinated operation. Having conquered Tibet, Chinese forces held the high ground. The PLA also gained a distinct advantage from its intensive preparations, which included building all-weather roads in Tibet to the Indian border, allowing the Chinese to move any heavy vehicle or heavy weapons system right up to the front. The Indians, by contrast, did not even suspect the PRC's territorial ambitions until they read of the encroachment in Chinese newspapers, and consequently they had made no preparations.

Without warning, just before dawn on October 20, 1962, the PLA simultaneously attacked Indian positions on the eastern and western sectors of the border. The heavy artillery and waves of infantry had overwhelmed the Indians by 9 AM.[28] Two Chinese army corps that had seen heavy combat in Korea did most of the fighting.[29]

The PLA's victory was complete. The Indians lost more than three thousand soldiers, while not a single PLA soldier was even captured.[30] Tens of thousands of square miles of Indian territory became part of China.

During the 1950s the PRC had expended a great deal of effort trying to make friends with India; the PRC's premier even made a triumphant tour of India in 1956. So, why a surprise attack on another innocent neighbor?

Given Chinese Communist policy toward India since 1962, it seems likely that Beijing wanted to prevent India from becoming a major world force. There was not going to be any "Third Way" (led by India) between Communism and the West.

If this was indeed the aim, the Chinese Communists were entirely successful. After 1962 India never had a major role in world affairs.[31] But China's nuclear and missile tests in the 1960s, combined with its ever increasing arms sales to Pakistan, have turned the Indian subcontinent into a nuclear armed camp.

Perhaps the saddest commentary comes from His Holiness the Dalai Lama, who watched Indian Prime Minister Nehru decline physically after 1962: "Many people say the Sino-Indian War broke his spirit. I think they may be right."[32]

THE SOVIET UNION, 1969

On March 2, 1969, the PLA and soldiers of the Soviet Red Army clashed on the northern border between China and the Soviet Far East. The world held its breath because this was the first direct fighting, ever, between two declared nuclear powers.[33] A second and heavier battle took place on March 15, but after that the opposing forces stood down.[34]

Each side promptly blamed the other for initiating the conflict, but the PRC reacted with an immediate and massive propaganda barrage that suggested prior planning. Beijing held anti-Soviet rallies around China involving some 260 million people within five days of the initial clash.[35] Meanwhile, Moscow reacted with confusion. By one high-level Soviet observer's account, the conflict was like an "electric shock" in Moscow; the Soviet Politburo had visions of millions of Chinese pouring into the underpopulated parts of Soviet Siberia and the Far East.

Professor John Garver points out, "There is a consensus among Western analysts that China took the initiative in the March 2 clash."[36] And most experts agree that the Chinese orders came straight from the top—probably from Mao himself.

VIETNAM, 1979

At 5 AM on February 17, 1979, the PLA launched a full-scale military assault across China's southern border with Vietnam. The highly coordinated surprise attack—involving 360,000 PLA soldiers, 1,000 tanks, and 1,500 pieces of heavy artillery[37]—struck in twenty-six places along the border, achieving complete tactical surprise and suggesting that the PLA had been planning the invasion for years.[38]

By the time the PLA withdrew to China a month later, it had left a twenty- to thirty-mile swath of complete devastation along the Vietnamese side of the border. The PLA had undertaken a campaign of deliberate destruction mostly aimed at the civilian economy; not a single light pole was left standing, and even hospitals were blown up.[39]

What was Beijing's motivation in launching a surprise attack on yet another neighboring country? As with its invasion of India in 1962, the answer is simple: regional dominance. The Chinese leadership stated publicly that it intended to "teach the Vietnamese Communists a lesson."[40] Before the invasion, the Vietnamese, claiming victory over the Americans, were a threat to the PRC's regional control, but the PLA's surprise attack made it apparent that Vietnam's Big Brother, the Soviet Union, was not willing to come to its aid in a time of crisis.[41]

The 1979 attack also allowed Beijing to test its current state of combat readiness. In fact, the PLA did not do particularly well against what were essentially second-line Vietnamese troops,[42] which served as a wake-up call to the Communist leadership. To some extent, the twenty-year PLA modernization program dates from the lessons learned in Vietnam.

LONG LIVE PEOPLE'S WAR[43]

The Chinese Communists did not limit their aggression to neighboring countries such as Tibet, Korea, India, Vietnam, and the Soviet Union. Beijing used armed subversion as a weapon against those nations of Southeast Asia the PLA could not reach on foot. Subversion was also used when deniability was an important consideration. Communist parties in Southeast Asia received guns and military training at PLA bases in China as well as secret financial and other support.[44] The war went on by other means.

THE PHILIPPINES

During the late 1940s and into the 1950s a Communist-led insurgency group known as the "Huks" plagued the Philippines. By the mid-1950s Chinese influence and support were heavy, and in the 1960s and 1970s lots of small arms and ammunition were smuggled to Communist rebels.[45] The democratically elected Philippine government is still fighting a Communist insurgency, and Beijing is still secretly supporting it.

MALAYSIA-SINGAPORE

Beginning in the late 1940s the British struggled against Communist Terrorists, influenced and supported by Beijing, in what is now Malaysia and Singapore.[46] Known as "the Emergency," the struggle took eleven years before the Communist Terrorists were finally defeated. Among the Communist Terrorists to die in jungle fighting at the hands of the British was Willy Kuok, the brother of Malaysian billionaire Robert Kuok, owner of Asia's Shangri-La luxury hotel chain and the *South China Morning Post*, Hong Kong's leading English-language newspaper.[47]

THAILAND

In the 1970s Beijing was supporting the Communist Party of Thailand with PLA-run "training facilities in south China, a clandestine radio station based in south China, political support, and (probably) arms and cash," according to Professor John Garver.[48] Thai authorities, led by the king himself, patiently improved living conditions in the countryside and destroyed the Communists' base of operations. By 1985 the terrorists were finished.

INDONESIA

Indonesia is Southeast Asia's most populous country and is rich in natural resources—oil, gas, rubber, timber. Consequently it became a prize worth fighting over.[49]

The crisis came to a head in 1965. In January the PRC premier secretly offered Indonesian supporters 100,000 PLA guns, and over the ensuing months clandestine shipments of Chinese arms began to arrive by sea, mostly at night.[50] On September 30, 1965, the Indonesian Communist Party launched a bloody coup d'état, murdering a number of the top Indonesian military officers.[51] General Suharto and his defense minister escaped, however, and the army and local people seized control from the Communists and killed perhaps half a million Communist Party members and ethnic Chinese—some guilty, some not.[52]

BURMA

On Christmas Day, 1968, thousands of Chinese "volunteer" soldiers streamed over the border into Burma. The Burmese army quickly found itself outgunned and outmanned as the Chinese infantry joined up with guerrillas from the Burmese Communist Party. According to one experienced observer, "The Chinese poured more aid into the Burmese Communist Party effort than any other Communist movement outside of Indochina."[53] The PLA gave the Burmese Communists mortars, recoilless rifles, machine guns, anti-aircraft guns, uniforms, and communications gear, and the PRC set up a "People's Voice of Burma" propaganda radio station in southern China. After six years of fighting, the Burmese Communists and their Chinese "volunteers" were able to carve out an enclave of twenty thousand square kilometers of Burmese territory next to the Chinese border.

> **Communist China's insatiable territorial aggression has claimed millions of lives.**

The Burmese Communist Party fell in 1989.[54] As we will see, Beijing and the Burmese military—the same military that stood up to the Chinese soldiers in 1968—are now comrades in arms against the tide of democracy in Asia.

VIETNAM AND CAMBODIA (INDOCHINA)

The CCP for years tried to manipulate Indochinese Communist parties—not always successfully. Immediately after taking over in China, the Chinese Communists began making plans to help the Vietnamese Communists grab power; Vietnamese Communist leader Ho Chi Minh had been a member of the PLA in the 1930s.[55] The PLA set up three training bases in southern China for tens of thousands of Vietnamese Communist recruits and funneled arms to them.[56] Without Chinese assistance, Ho could not have defeated the French at Dienbienphu.[57]

According to the Chinese themselves, between 1964 and 1971 they provided North Vietnam with more than 2 million rifles, more than a quarter of a billion rounds of ammunition, 37,000 artillery pieces, and almost 20 million artillery shells.[58] The PLA's war materials were undoubtedly responsible for thousands of American casualties during the Vietnam War, and Chinese anti-aircraft batteries stationed in North Vietnam probably accounted for some American air losses.

The PRC has also supported Cambodia's Communists, the notorious Khmer Rouge. Armed with an assortment of PLA and Soviet-bloc weapons,[59] the Khmer Rouge swept to power in 1975. By one expert estimate, a Communist victory in Cambodia cost the lives of two million persons out of a total population of perhaps ten million.[60]

A RECORD OF AGGRESSION

The PLA's invasions and armed subversions of neighboring countries have wrought destruction throughout Asia. The Chinese Communists' brutal territorial aggression in the past fifty years has led to the deaths of:

- More than fifty thousand Americans in Korea
- More than fifty thousand Americans in Vietnam
- One million Tibetans
- One million Koreans
- Two million Cambodians
- One million Vietnamese

Former Secretary of State Henry Kissinger and other Western political figures contend that the PLA "is not a threat."[61] But a review of the historical record reveals the danger in ignoring the significant threat the PLA poses—a threat that is already recognized in democratic capitals from Tokyo to New Delhi.

PART II
UNCHECKED
PROLIFERATION

CHAPTER 5
DRUGS AND THUGS

"Keep the Profits"

—The name of the PLA's leading arms smuggling company

G iven his choice, Undersecretary of State Reginald Bartholomew would probably have preferred to be somewhere else, but an urgent June 1991 trip to Beijing was part of his responsibilities.[1] Three months before, in March 1991, the *Washington Times*'s Bill Gertz had reported that the Chinese had delivered to Pakistan equipment necessary to help develop its nuclear-capable M-11 missile system, and a bipartisan Capitol Hill posse led by Senate Majority Leader George Mitchell (D-Maine) and Senator Jesse Helms (R-North Carolina) was now demanding sanctions against the offending PLA-associated exporters.[2]

The undersecretary was looking for a no-more-missile-exports commitment from the right Chinese official so the administration would not sanction Chinese arms companies. For years the Chinese Foreign Ministry had been avoiding American complaints about proliferation by pointing to the PLA as the ones in authority. After a good deal of persistence,

Bartholomew obtained an appointment with General Liu
Huaqing, then one of the PLA's highest ranking officers, but by
all accounts the meeting did not go well.[3] Within a month, the
Bush administration sanctioned two Chinese arms companies,
Great Wall Corporation and China National Precision Machinery Import-Export Corporation, for M-11 missile–related
exports to Pakistan.

We now know that General Liu's daughter, Lieutenant
Colonel Liu Chaoying, is both a PLA spy and an arms dealer.
In fact, during her career, she has been "assistant to the president" of each of the sanctioned companies.[4] The link is significant; as former Secretary of State James Baker revealed in his
memoirs, key PLA officers and their families hold lucrative missile contracts. As a practical matter, by asking the PLA to halt
missile exports, the undersecretary was asking General Liu to
smash the Liu family rice bowl.

Lieutenant Colonel Liu is also a member of the Er Bu, the
PLA's military intelligence department.[5] In recent years her
immediate boss has been General Ji Shengde, the man who,
according to Democratic National Committee donor Johnny
Chung's federal grand jury testimony, gave Chung $300,000 to
funnel into President Clinton's 1996 reelection campaign. After
Chung's testimony, Ji was quietly given a lateral transfer out of
the public eye. Chung, for his part, was convicted of making
hundreds of thousands of dollars in illegal campaign contributions in 1996 to the Democratic National Committee.[6]

What follows is the story of the Chinese arms companies,
the Communist aristocracy that dominates many of them, and
the threat they pose to Americans and to our democratic allies.
Unfortunately, it is a threat made worse when Chinese government officials and arms dealers can get easy access to the president of the United States.

THE COMPANIES

Raging over eight years, the Iran-Iraq War resulted in almost a million soldiers killed or permanently disabled. And each year of the war, Chinese arms companies racked up billions of dollars selling weapons *to both sides*. In the summer of 1987 their international banker noticed an interesting connection. Coming across his desk for routine trade finance from one PLA-associated arms company was a sale of sandbags to Iraq—to counter the sale of munitions the company had just made to Iran.[7]

The arms companies have names that sound exotic to the non-Chinese ear—"Blue Sky Industrial Corporation," "Keep the Profits, Inc.," "Sky Horse Enterprises," "Carrie Enterprises," "Rainbow International," "Great Wall Industries," "999." All are well known in the global arms bazaar that has continued unabated since the end of the Cold War. They produce, broker, sell, transport, and otherwise accommodate the trade in ballistic and cruise missiles, nuclear weapons technology, germ warfare, poison gas, tanks, artillery, anti-personnel mines, laser-blinding devices, aircraft, ships, rockets, shells, and infantry weapons of all kinds. Their customers comprise the most loathsome dictators in the world: the military junta in Burma as well as all the terrorist regimes—Iran, Iraq, Syria, Libya, North Korea, Cuba, and Sudan.

The arms companies can be roughly divided into three groups:

- PLA-owned companies
- PLA-associated companies
- Auxiliaries

The PLA-owned companies: The PLA entered into business in the early 1980s with Polytechnologies (now part of Poly Group), a major missile broker whose Chinese name, *Baoli*, means "keep the profits." Poly's success soon drew competitors and copycats. By the middle of 1998 each of the four PLA departments[8] had commercial companies, as did the service branches, all seven military regions, and the People's Armed Police.

The PLA-associated companies: These nominally civilian firms produce everything from uniforms to missiles for the PLA. They differ from defense contractors in democratic countries in that the Chinese firms are all government-owned. North China Industries (Norinco) claims to be the largest arms manufacturer in the world, with almost a million employees. China Aerospace Corporation is another major player. The PLA arms dealer/spy Lieutenant Colonel Liu Chaoying's employers are subsidiaries of China Aerospace.[9]

The PLA-associated firms are arms producers while the PLA-owned companies are essentially brokers, and the producers would certainly like to get rid of the middle man to increase profits. Nevertheless, the two do work closely together on major projects. For example, the two thousand machine guns imported into the United States in 1996 for sale to youth gangs was a joint Poly-Norinco operation. Likewise, Poly brokers missile sales to the Middle East for China National Precision Machinery.

The auxiliaries: China Ocean Shipping Company (COSCO), delivery boys for the arms dealers,[10] and China International Trust and Investment Corporation (CITIC), a ministerial-level Chinese firm, are both auxiliaries of the PLA and PLA-associated firms. For the first several years of its life, Poly received cover as a supposed subsidiary of CITIC, and its PLA officers traveled overseas in civilian clothes as CITIC employees. Today, the chairman of CITIC is also the chairman of Poly.

By some accounts the PLA's business empire comprises 20,000 companies, employing 600,000 people and having a turnover of at least $20 billion per year. By the summer of 1998, the *Washington Post* reported, it "controlled 20 percent of China's automotive industry and ran nearly 400 pharmaceutical factories, 1,500 hotels and four of China's 10 biggest clothing factories."[11]

In theory, all this has come to an end. In July 1998 President Jiang Zemin announced that all military companies would be closed by the end of the year. It's probably true that some PLA money-losers have been dumped, but the profitable ones, especially the arms smugglers, will remain. From time to time the PRC press announces that this or that PLA unit has turned over its commercial activities to local authorities. In March 1999, for example, there was a lot of publicity as the

Leading Chinese Communist families line their own pockets with the profits from arms sales to terrorist regimes.

five-star Palace Hotel in Beijing was turned over to the Everbright Group, a major government-owned conglomerate.[12] According to a well-connected press source, however, the second in command of Everbright's Hong Kong office is the chief military intelligence officer in Hong Kong.[13]

The PLA companies, the PLA associates, and the auxiliaries have two major missions: (1) sell products, including civilian products, abroad, and (2) acquire foreign technology to assist the PLA's modernization efforts. The PLA's so-called "sixteen-character" policy[14] lays out the objectives:

- "Military-civilian unity"
- "Peacetime-wartime unity"
- "Priority for military production"
- "Use civilian production to support the military"

America's Defense Intelligence Agency explains the sixteen-character policy as follows:

The "sixteen-character policy" serves China's military establishment in several ways. For example, profits from the sale of mostly low-tech, low-cost consumer goods pay for the development or production of new weapons.

Second, those few military industries producing sophisticated civil products (airliners and commercial ships) can and do use them as a cover for acquiring or adapting dual-use technologies (guidance systems, machine tools, and computers) for weapon programs at those facilities.

Also, in some instances, production of civilian products temporarily takes the place of deferred weapons such as tanks or armored personnel carriers until China's leaders determine they are needed.

Finally, as a recent twist on the policy, China seeks foreign investment in joint ventures to modernize the military-industrial complex. In many instances, Western investors are establishing joint ventures with facilities that produce cruise missiles, guidance systems, and electronic warfare equipment.[15]

In practical terms, this means that:

- a substantial amount of money comes to the PLA from arms sales abroad;[16]
- ordinary civilian products you may purchase at Wal-Mart, for example, may have been made in a PLA-owned or associated factory;[17]
- if you travel on China United Airlines, your host is the PLA air force;

- multinational companies are deeply engaged in joint ventures with PLA companies in China, and the net effect may be an enhanced PLA warfighting capability; and
- America's commercial ties with PLA firms may be a leading conduit for espionage.[18]

Arms sales are the biggest profit center. Poly made $2 billion in net profits on a single missile sale to the Middle East.[19] But PLA companies and their associated companies make everything from refrigerators to automobiles to rubber boots—anything that can help them turn a profit.

How much money are we talking about? Lots. According to their own figures, PLA companies made $7 billion in 1997,[20] and most outside observers believe the companies keep more than one set of books.

So, who gets the money?

THE PRINCELINGS

There's no question that the profits from arms sales are huge. What outsiders can only speculate about is how the pie is divided. Even while the PLA looks to use the profits to modernize its weaponry, its leading families are looking to line their own pockets.

The major figures who brought the PRC into existence fifty years ago have mostly passed from the scene, but in their wake they left a Mafia-like political structure dominated by "princelings"—sons and daughters of former Communist Party officials. Notable princelings include the president of the PRC, Jiang Zemin;[21] Li Peng, the head of China's rubber-stamp parliament;[22] and many of the advisers who surround the PRC's premier.[23] Trying to figure out who is really the son or daughter of which official is a common parlor game in China.

The PLA and its related organizations are so infested with princelings that some Hong Kong publications refer to the PLA as "The Princes' Army."[24] Lieutenant Colonel Liu Chaoying is just one example. Her boss, General Ji, is himself the son of a former official,[25] as is one of Ji's subordinates.[26] One of Deng Xiaoping's daughters is an official of the PLA's General Political Department.

The princelings know that the PLA's arms companies are where the money is. According to one London arms dealer, "The lucrative arms business is a very attractive niche for the princes.... They [the arms companies] are run and operated by and for the benefit of the senior brass and their families."[27] The princelings use their unique relationship to their parent or in-law to gain access to the latest PLA weaponry and the requisite export permits. In some cases the most modern arms have been exported, leaving the PLA with inferior equipment.[28]

> **The Chinese Communists have no problem trading with Burma's brutal military junta—anything to turn a profit.**

Poly is the most notorious haven for princelings. Its chairman is Wang Jun, son of the late hard-liner General Wang Zhen. The father was instrumental in the massacre of Chinese young people at Tiananmen Square, and the son was a source of embarrassment to the Clinton administration when the *Washington Post* revealed that he had made an appearance at a White House coffee/fund-raiser.[29] The president of Poly is one of Deng's sons-in-law.

This is no accident, as an arms dealer emphatically told one of the authors.[30] Standing in the elegant drawing room of his multimillion-dollar Hong Kong flat, the arms dealer explained how Communist Party officials divide up the pie. The PLA and the Party, he said, operate like a mafia family. Communist Party officials assign their family members to Poly's various companies as a New York City godfather would assign, say, control of a Brooklyn racketeering operation. The primary difference here is that local mafia families are generally not dealing in weapons of mass destruction.

Other PLA arms companies such as Carrie Enterprises have a history similar to Poly's.[31] For years the Commission on Science Technology and Industry for National Defense (COSTIND), China's nuclear weapons makers, was dominated by one military family.[32]

So, just how much of the profits goes into the pockets of these clans? Based on sensitive conversations in Hong Kong and elsewhere, we believe that the princelings have made hundreds of millions of dollars, if not more, on the arms trade in the past twenty years. Family members can draw on these funds for foreign travel, overseas education, and luxury goods such as automobiles.

The prime function of these ill-gotten gains is to serve as an overseas secret stash in the event that Communism ends in China. Since the fall of the Berlin Wall in 1989, the Chinese Communists have seen many of their old East European comrades impoverished.[33] And as one Hong Kong arms dealer noted, "They know they are hated by the Chinese people and retribution is certainly likely."[34]

It's no surprise that the Chinese people resent the privileges bestowed on the princelings. These elites have their own middle schools and have had access to study abroad, first in the Soviet Union and, more recently, in the United States. Some princeling children are now left with relatives or paid sponsors in the United States. General Ji Shengde's son has been in the United States for ten years and speaks English like a native.[35] President Jiang Zemin's son has a United States green card, and

Deng Xiaoping's grandson was born in America, according to a spokesman for Hong Kong's Information Centre of Human Rights and Democratic Movement in China.[36]

The clans' ostentatious displays of wealth and power make no secret of their status. "The princelings," reported Hong Kong's *South China Morning Post*, "cruise around town in limousines with PLA or PAP [People's Armed Police] license plates and entertain in exclusive clubs that are guarded by their own security."[37] According to former U.S. Ambassador to China James Lilley, the princelings "like golden Rolls-Royces, wet 'T' shirts and foreign bank accounts."[38] Professor June Teufel Dreyer notes that the princelings have access to "foreign travel, expensive consumer items and the good life in general."[39] As a result of such behavior, nepotism was one of the major complaints of Tiananmen Square protestors.

FUNDING THE HIGH LIFE

A summer 1988 telephone call from Washington to the deputy chief of mission at the United States Embassy in Rangoon, Burma, went something like this:

> *Question*: What's that noise in the background? I can barely hear you.
> *Answer*: It's the Burmese students marching past the embassy, carrying the American flag and cheering![40]

For their devotion to the cause of democracy, the Burmese students were machine-gunned by their own military on August 8, 1988. Thousands died.[41] Since then, Burma has been ruled by a military junta, sometimes called the "State Law and Order Council," or SLORC.[42]

Because of the illegitimacy of the regime, the United States declines to send an ambassador to Burma and in 1997 imposed economic sanctions.

But the Chinese Communists have no problem trading with this brutal military junta—anything to finance the princelings' lavish lifestyles and the PLA's extensive military buildup.

The Burmese people, led by 1991 Nobel Peace Prize–winner Aung San Suu Kyi,[43] are willing to rebuild their country and be a peaceful presence in Southeast Asia, but the morally and financially bankrupt military junta has thus far kept down the democracy movement. And Beijing has played a vital role in keeping this abominable regime[44] in power. In the past ten years, Beijing has sold the junta almost $2 billion worth of arms. These deals have been brokered by Poly, and the arms have been produced by Norinco and delivered by COSCO.[45] Among the arms sold to the SLORC are:[46]

> **Because of Communist China's unchecked weapons proliferations, *every* country is at risk of annihilation—including the United States.**

- F-6 jet fighters
- F-7 jet fighters
- Armored personnel carriers
- T-63 tanks
- Anti-aircraft guns
- Artillery
- Radar and signals intelligence (SIGINT) equipment
- Rocket launchers
- Surface-to-air missiles
- Light arms and ammunition
- Five-ton trucks
- Naval frigates and patrol boats

Armed with Chinese weapons, the Burmese military grew from 170,000 men to 300,000 by 1995 and may reach 475,000 by the year 2000.[47] At this rate, it will have 50 percent more soldiers than neighboring Thailand and will easily be able to suppress dissent from unarmed civilians.

But the SLORC has driven the civilian economy into the ground, which raises the question, how is it paying for this fancy military hardware? The answer, it seems, is narcotics.

The vast majority of the heroin sold "on the streets of the U.S. originated in Burma," in the words of one U.S. government official.[48] From time to time the SLORC has made a big show of burning some confiscated drugs, but in reality the military rulers are narco-traffickers who have driven their country to ruin and sold the remains to Beijing.[49] Since the regime took power in 1988, heroin production has doubled; Assistant Secretary of State for International Narcotics Affairs Robert Gelbard called the SLORC "part of the problem, not the solution."[50]

But is the PRC also involved? In 1992 Senator Paul Simon (D-Illinois), a senior member of the United States Senate Foreign Relations Committee, writing in an Institute for Asian Democracy publication, stated flatly, "Certainly, some of the proceeds [from heroin sales] are going to purchase Chinese military hardware, an estimated $1 billion in recent years."[51] The Institute for Asian Democracy report made the direct drugs-and-guns connection to Poly and Deng Xiaoping's son-in-law, a charge that, to our knowledge, has never been denied.[52] Whatever applies to Deng's son-in-law may apply to Poly Chairman Wang Jun, as well.

In 1997 a Hong Kong newspaper ran an explosive story accusing another Politburo member's family of being engaged in drug smuggling with Burma.[53] And witnesses claim that one night in the mid-1990s, in a Burmese border village, a line of

Burmese army trucks lined up tailgate to tailgate with a corresponding line of PLA trucks. According to the witnesses, the Burmese soldiers gave heroin to the PLA in exchange for arms.[54]

We believe it is impossible for Beijing's leadership at the highest levels not to know how the SLORC pays for these weapons. Everyone is complicit.

PLA, INC.

As we will see, the Burmese military junta is not the only brutal regime to which the PLA and its companies are willing to sell arms. Chinese arms dealers are willing to trade any weapon with anyone at any time, as long as the price is right. All the while, the obscene profits from this trade go to the overseas bank accounts of the very same families who suppress democracy in China.

Because of Communist China's unchecked weapons proliferations, *every* nation is at risk of annihilation—including the United States and its allies throughout Europe, the Middle East, and Asia. In most cases, Chinese arms deals involve weapons of mass destruction[55] or the ballistic and cruise missiles to deliver them. In other words, we are talking about nuclear war.

CHAPTER 6
A PRESENT FOR SADDAM

There is nothing more important to our security and to the world's stability than preventing the spread of nuclear weapons and ballistic missiles.[1]

—President William J. Clinton

To Group Captain William Cross, Britain's military attaché in Algeria, the 1991 assignment must have seemed like something out of a James Bond film. His mission: Proceed to a secret nuclear weapons site being built in the Sahara Desert about eighty miles south of Algiers, take close-up photographs, and return without being discovered.

The facility in question lay within a military exclusion zone at the foot of the Atlas Mountains. The entire complex was surrounded by a double set of high-security fences. Just outside the outer fence was a carefully plowed killing zone of anti-personnel mines and electronic sensors. Inside the first perimeter the facility was divided into eight security zones, each protected by fencing, multiple check points, and armed guards.[2] Satellite photography showed that nearby was a military airfield with two hardened runways, one 3,200 meters long—more than adequate for heavy-lift transports—and another 2,600 meters—appropriate for modern jet-fighters. Soviet-produced SA-5 surface-to-air missiles with a range of 150 miles provided an outer ring of protection.[3] With this level of security, the Algerian military was prepared for any eventuality—an air attack, a serious commando ground assault, or intruders on an espionage mission.

Under these truly formidable circumstances, Captain Cross had no chance. He was captured and expelled from the country. The British Foreign Office quietly reported that he had been recalled "for operational reasons."[4]

The Algerians had clearly gone to great lengths to guard the facility. But what, exactly, were these extraordinary precautions designed to protect? The six large cooling towers, the lack of electrical generating equipment or transmission lines, the secrecy under which the facility was being built, and the extreme military security all pointed in one direction. Based on satellite inspection, allied intelligence had come to a startling conclusion: A Chinese government arms maker was building nothing less than a clandestine nuclear weapons facility in the Algerian desert. According to French intelligence, "It is simply not feasible to use this reactor for civilian purposes. You have to conclude that it is for military purposes."[5] Based on its size, allied specialists estimated that at the very least the facility could produce, every three years, the nuclear materials[6] for two bombs the size of the bomb dropped on Nagasaki. The Chinese arms makers were moving at a rapid pace, and the first bomb-grade materials would have been ready by 1995.[7]

But China was not merely producing bomb-grade nuclear materials. Early in its investigation American intelligence had concluded that Chinese scientists were "supplying nuclear-weapons technology and military advice on how to match nuclear weapons to various aerial and missile delivery systems."[8] Later, British intelligence prepared a top-secret report for the British cabinet accusing the Chinese of transferring "warhead design technology"—that is, the triggering mechanism for missiles—to the Algerian project.[9]

Within twenty-four hours of Captain Cross's safe return to England, the *Washington Times* ran this headline: "China helps Algeria develop nuclear weapons."[10] So began an intense ten-month battle between allied intelligence officers and diplomats on one side and Chinese and Algerian government officials on the other. In the course of it, at least one Algerian agent for the British would be captured, tortured, and executed;[11] the Algerian government would come under overwhelming political and economic pressure from abroad; and the Chinese government would be denounced as a "rogue elephant" on the floor of the Senate by Senator Joseph Biden (D-Delaware).[12]

As soon as the *Washington Times* broke the story, both Algeria and China began to stonewall. The Algerian official media referred to the Chinese reactor as "imaginary" and Captain Cross's report as "fantasies."[13] For the first three weeks the Chinese denied everything in public and even in private diplomatic consultations with highly skeptical allied governments. But the project was too big to hide, and they knew it. By the end of April 1991, a coordinated Algerian-Chinese diplomatic offensive agreed that the project existed but that it was "entirely for peaceful purposes," and a Chinese spokesman claimed the facility would eventually be put under international inspection[14]—though it would be another eight months before the first inspectors from the International Atomic Energy Agency (IAEA) visited the plant.

But the Chinese-Algerian stonewalling was not the most troubling aspect of the case. The Algerian government's decision to embark on this project seemed mysterious. The Algerians were already negotiating with Argentina for a small, one-megawatt reactor that would be more than adequate for Algeria's legitimate research needs. Nor was there any sort of external security threat to Algeria sufficient to justify a project as expensive as this; even neighboring Libya did not pose a substantial threat, since the tough Algerian military could overwhelm Libyan leader Muammar Qaddafi's much smaller armed forces.[15] Moreover, Algeria could not justify a nuclear reactor on the grounds of energy needs; the OPEC member nation has the fifth largest natural gas reserves in the world and ranks fourteenth in oil production.[16]

Even more puzzling, Algeria was in no position to take on such a financial burden. By the mid-1980s the socialist government's gross mismanagement had left the country with a crushing external debt and living standards so bad that serious rioting erupted in 1988. Nuclear weapons programs are incredibly expensive, and Algeria was so strapped for cash it was even calling in loans from economically desperate countries such as Vietnam.[17]

NATO analysts also had reason to pause because, even with nuclear materials and the triggering mechanism for missiles, the Algerians had no ballistic missile delivery system, not even Soviet-built Scuds.

So why had Algeria embarked on this venture? Because someone else was paying for it—someone who *could* launch missiles.

THE PLA HELPS SADDAM

Flash back ten years earlier to 1981, not Algiers or Beijing, but Baghdad, the capital of Iraq. On the evening of June 7, 1981, Israeli pilots flying American-made F-16s dropped a series of bombs on a forty-megawatt research reactor under construction by the French. Within moments, Saddam Hussein's dream of an

Iraqi bomb, and his investment (reputed to be $750 million), went up in smoke.[18]

From this setback, the Iraqis decided that any future advanced weapons programs would be based on two principles: *deception* and *redundancy*. They went to elaborate lengths to disguise buildings and other weapons facilities so that they could not be identified or targeted from the air. This was enormously expensive, but they felt it was necessary to create duplicates and multiple back-ups so that the loss of

Thanks to the Chinese, Saddam Hussein could have enlisted nuclear weapons to support his territorial ambitions.

one element would not jeopardize the entire program. There was not a single nuclear program or a single missile program but rather multiple nuclear and missile programs operating simultaneously.

Saddam also needed to replace the bombed-out reactor. But after the Israeli bombing experience, the Iraqis knew that a new nuclear reactor would have to be hidden carefully. According to a now-declassified report prepared for the United States Defense Nuclear Agency, "Between 1982 and 1986, Iraq opened three sets of discussions on the purchase of an underground nuclear reactor: with the Soviet Union, a French-Belgian consortium, and China."[19] A 1986 U.S. Army intelligence report indicates that the Iraqis were insisting that the underground reactor be defendable "from possible attacks" and that it should have an "ability to camouflage from satellites."[20]

After the 1990–1991 Gulf War, United Nations inspectors in Iraq found the negotiation documents on the underground reactor but not the facility itself, leading some to speculate that the "facility has not yet been discovered."[21]

The reason the facility was not discovered is, we believe, simple: *Saddam's nuclear weapons reactor was indeed constructed, but above ground in the Algerian desert, not underground in Iraq.*

A major nuclear reactor is extremely difficult to hide. The large cooling towers cannot be disguised, and its heat signature is easily visible from space. Even China's secret reactor project in a quiet corner of the Sahara Desert of Algeria was discovered long before it became operational. Given the intense and sophisticated intelligence scrutiny Iraq was under during the 1980–1988 Iran-Iraq War, a nuclear reactor, even underground, would have been uncovered quickly. And once discovered, it would have been destroyed.

The Chinese had their own motivation to build in the Sahara. PLA companies, principally Norinco and Poly, were making enormous profits selling to both sides in the Iran-Iraq War—almost $8 billion, according to U.S. congressional specialists.[22] To be found building a nuclear weapons plant of that magnitude in Iraq would likely have jeopardized their war goods market in Tehran. From China's standpoint, it would be far better to build the reactor far from Iraq, where it would be less likely to be discovered and, if discovered, could feasibly be denied.

It was also reasonable for the Algerians to help Saddam, since the connection between Baghdad and Algiers was very strong. During the Gulf War hundreds of thousands of people enthusiastically demonstrated for Saddam in the streets of Algiers and other major Algerian cities. Within two weeks of the *Washington Times* story on the Chinese nuclear plant, a prominent Arab newspaper claimed that the facility had been "partly financed by Iraq"[23]—no surprise, given the Algerian government's difficult financial situation. Later, American intelligence received reports that Iraqi nuclear scientists were working alongside the Chinese in Algeria on sensitive nuclear weapons technology.[24]

After Iraq's August 1990 invasion of Kuwait, Algeria proved to be a willing co-conspirator with Iraq in an elaborate effort to keep Saddam's Iraq-based nuclear weapons program away from United Nations inspectors. French intelligence reported that much of Iraq's nuclear equipment and material was secretly flown to Algeria a few days after the United Nations Security Council passed its first resolution imposing a naval and land blockade,[25] and another significant shipment went to Algeria in early May 1991, just before the first United Nations inspectors were to arrive in Iraq. Britain's supersecret eavesdropping organization, the Government Communications Headquarters (GCHQ) at Cheltenham, intercepted crucial telephone conversations among senior Iraqi leaders in Baghdad discussing ten tons of Iraqi natural uranium that had been sent to Algiers. This uranium transfer occurred at precisely the same time Chinese and Algerian officials were trying to divert allied attention from the reactor in the Sahara. With the Iraqi natural uranium in hand, the Algerian reactor would have enough fuel to run well into the twenty-first century.

ARMED AND READY

A nuclear weapon is, of course, useless without a delivery system, and the Chinese were just as willing to help Iraq develop its missiles. A full year before they found the Chinese reactor in Algeria, American satellite reconnaissance specialists had found and identified Saddam's missile base in an even more remote corner of the Sahara—Mauritania.[26]

The Central Intelligence Agency (CIA) describes the Islamic Republic of Mauritania as "desert, constantly hot, dry, dusty, sparsely populated"[27]—an ideal place to hide a missile test range and related equipment. Increasing desertification in the late 1980s combined with economic mismanagement left the

Mauritanian government with a heavy foreign debt load similar
to Algeria's,[28] and Saddam's financial inducements proved to
be as attractive to Mauritania's rulers as they had been to the
Algerian government.

In the Mauritanian missile base, allied intelligence had discovered Iraq's grand scheme to create, in total secrecy and with
the willful aid of the Chinese, a complete Iraqi offshore strategic
weapons system—bomb-grade nuclear materials, warhead
design and assembly, and a long-range ballistic missile. At least
two-thirds of the system—nuclear materi-
als and warhead development—should be
stamped "Made in China." This was Saddam's ace-in-the-hole, complementing all
his in-country strategic weapons programs.

**Chinese arms dealers have
transferred nuclear, chemical, and biological wea-
pons to the most depraved
tyrannies of the late twentieth century.**

The security threat from a long-range,
Iraqi missile-delivered nuclear weapons
system would have been massive. Saddam
was known to be developing a 1,200-mile missile.[29] From Iraqi
territory Saddam would have been in position to threaten all the
oil-rich countries of the Middle East (and Israel) with nuclear
annihilation. Based in Algeria, such a weapons system could
have reached Paris. Had his plans not been stopped, Saddam
would have enlisted nuclear weapons to support his territorial
ambitions.

And the PLA arms companies were in it up to their necks.[30]

In the case of Iraq's indigenous arms programs, Chinese companies have been active for decades, especially in the nuclear
program. The IAEA reports that sometime in the 1970s Iraq
came into possession of almost two tons of enriched uranium
recovered from a Chinese military reactor.[31] By the 1980s Iraqi
scientists had turned to high-speed centrifuges as one method of
producing bomb-grade material, and for a fee (presumably

large) the PRC clandestinely supplied Iraq with the special samarium-cobalt magnets that are critical to the centrifugal process and are extremely difficult to obtain or manufacture.[32] This sale remains so politically sensitive that the United States Defense Intelligence Agency has refused Freedom of Information Act requests for *any* information in its possession.[33] After the Gulf War, United Nations inspectors inventorying Iraq's nuclear weapons program found other "sensitive" Chinese equipment being used for uranium enrichment and actual weapons detonation.[34]

At the time of the invasion of Kuwait, August 2, 1990, Chinese arms companies were deeply engaged in Iraq. Wanbao Engineering, an engineering and construction arm of Norinco, had more than five thousand Chinese nationals working on important projects, some of them military-related.[35] Three days after the invasion the Chinese government joined the United Nations arms embargo on Iraq, but allied intelligence soon learned that Wanbao was secretly trying to sell Iraq seven tons of lithium hydride, a chemical compound essential to hydrogen bombs, missile fuel, and nerve gas.[36]

In spite of the international outcry over the lithium hydride deal, and in spite of the continuing arms embargo, China has never ceased trying to deliver military equipment to Iraq. In December 1990, on the eve of the air war in the Gulf, allied intelligence discovered Chinese leaders searching for third countries willing to serve as conduits for arms smuggling operations.[37] By late April 1991, after the end of the ground fighting, the Chinese found individuals in Singapore and Jordan willing to pass military spare parts and ammunition to Baghdad.[38] At a May 17, 1991, Senate Foreign Relations Committee hearing, Senator Alan Cranston (D-California) asked Assistant Secretary of State Richard Solomon about reports that China had secretly provided military supplies to Iraq during the war; Solomon was careful not to deny the reports and asked to reply at a closed-door session.[39] As recently as September 1994 German authorities confiscated a shipment of ammonium perchlorate (rocket fuel) bound for Iraq from the Chinese Chemical Import-Export Corporation.[40] United Nations inspectors who were responsible for dismantling Iraq's weapons programs in 1996 privately report that Chinese companies are still trying to evade the arms embargo by fulfilling pre–Gulf War contracts.[41]

The PRC has been careful to conceal its aid to Iraq's missile programs. For example, in the course of its investigation of the shadowy South African arms company Armscor, an official South African government commission discovered that in the late 1980s, under the apartheid regime, Norinco had used Armscor as a conduit to illicitly transfer American missile technology to Iraq.[42] In return, South Africa received Chinese long-range missile technology.[43] The Chinese may have incorporated the American missile technology they received from this complex deal into the Scud missile production facility they built near Baghdad.[44] This facility may have produced the extended-range Scud missile that hit a U.S. Army mess hall outside Dhahran, Saudi Arabia, during the Gulf War, killing twenty-eight Americans and wounding ninety-eight others—the most casualties the United States suffered in any single episode in the Gulf War.[45]

Similarly, the Chinese made a secret deal with the Brazilian aerospace firm Centro Technico Aeronautico (CTA). As part of its civilian space-launch program, CTA had legally obtained solid rocket propellent technology from the United States, and in 1986 it swapped this American technology for Chinese assistance on missile guidance and liquid rocket fuel. Fairly promptly, the U.S technology made its way to an Iraqi laboratory.[46]

ALGERIA REVISITED

With the Chinese nuclear plant in the Sahara under IAEA safe-guards and Saddam bottled up, one would think the Algerian problem would be solved. But Spanish intelligence warns that Algeria could be producing military plutonium by the middle of the year 2000.[47] For what purpose, no one knows. As one of America's leading think tanks, the RAND Corporation, notes dryly, "Many of the world's leading WMD [weapons of mass destruction] proliferators are arrayed along Europe's southern periphery, and WMD risks are transforming the security environment in the Mediterranean as well as in Europe's regions."[48] Every one of those proliferators is a Chinese customer.

In short, the danger of nuclear, chemical, and biological annihilation being spread by Chinese arms dealers is now everyone's problem. System after system has been transferred, totally in secret, to the most depraved tyrannies of the late twentieth century. Every democratic country in the world is at risk.

It is only a matter of time before one of these weapons is put to use.

CHAPTER 7
ARMING THE TERRORISTS

I believe the proliferation of weapons of mass destruction presents the gravest threat that the world has ever known.[1]

—Secretary of Defense William Cohen

During the last half of 1996, China was the most significant supplier of Weapons of Mass Destruction related goods and technology to foreign countries.[2]

—Central Intelligence Agency

T oday there exists a kind of worldwide guerrilla warfare between the allied intelligence services on the one side and the Chinese arms merchants and their vile customers on the other. The PLA-associated companies want to protect their very lucrative business, which generates about $7 billion per year.[3] The intelligence services know that success or failure can determine national survival.

In most cases, the battle is seldom as dramatic, or successful, as the Algerian example would lead one to expect. More often the intelligence services are groping in the dark. For example, according to the *New York Times*, allied intelligence officials in late 1990 believed that Iraq would produce its first nuclear weapons in the year 2000,[4] but in fact United Nations inspectors later concluded that, absent its defeat in the Gulf War, Iraq could have produced its first nuclear weapon in mid-1991.[5]

For obvious reasons, the Chinese government does not want to be known as the world's leading arms dealer. China conducts no open sales of weapons of mass destruction to Third World dictators, and, when arms deals are discovered, Chinese officials respond with denials, sometimes very emotionally. As a result, Chinese arms transfers are probably much more extensive than even the intelligence services suspect.[6]

Communist China's leadership has no hesitation in subverting American anti-proliferation goals, if the price is right.

But Beijing's denials notwithstanding, the CIA and Department of Defense recognize the threat posed by Chinese arms dealing. A highly classified October 2, 1996, CIA report entitled "Arms Transfers to State Sponsors of Terrorism" named the PRC as the world's leading proliferator,[7] and the Department of Defense has placed Chinese companies at the center of the spider's web of worldwide illicit arms transfers.

Of all the governments the U.S. State Department identifies as terrorist regimes—the governments of Iran, Iraq, Syria, Libya, Sudan, North Korea, and Cuba—every one is a prime customer of China's arms merchants.[8] It is clear that the Chinese have launched a massive campaign to arm the world's most dangerous regimes with weapons of mass destruction and their missile delivery systems, as well as the most advanced conventional weapons.

IRAN: A DIRECT THREAT

Chinese clandestine arms sales to Iran have been extensive—nuclear weapons technology, ballistic missile technology, chemical weapons, biological weapons, and advanced conventional weapons. The Chinese and Iranians have cooperated so much on arms deals that, it seems, the dealers have grown quite chummy. At the March 1999 International Defense Exhibition[9] in Abu Dhabi, a group of about a dozen Chinese arms dealers showed up at the Iranian arms merchants' booth—and *everybody knew everybody!*[10]

The American Office of Naval Intelligence compares China's weapons buildup of Iran to the situation in Iraq before 1991:

> Discoveries after the Gulf War clearly indicate that Iraq maintained an aggressive Weapons of Mass Destruction procurement program. A similar situation exists today in Iran with a steady flow of materials and technologies from China to Iran. This exchange is one of the most active WMD programs in the Third World, and is taking place in a region of great strategic interest to the United States.[11]

One reason Iran is a major threat is that it is a key sponsor of terrorist groups. The Iranian Revolutionary Guard is a well-established patron of the Hizballah guerrillas in Lebanon, sometimes using Syria as a transit point.[12] The Iranians have even been accused of conspiring with Chinese weapons dealers to ship chemical weapons to the radical regime in Sudan.[13]

But Iran's *nuclear* program is our first concern. American intelligence believes that Iran has an aggressive nuclear weapons program,[14] despite a string of Iranian denials.[15] General Binford Peay, commander-in-chief of the U.S. Central Command,[16] told reporters in the summer of 1997 that Iran will have an explosive device "in the near end of the turn of the century."[17]

Therefore the American government opposes any transfer of material or technology to Iran that could help establish a nuclear infrastructure. And since the late 1970s the United States has made well known its opposition to the training of Iranian nuclear specialists in foreign countries. For example, at a news conference on April 17, 1995, Secretary of State Warren Christopher said that Iran "is simply too dangerous with its intentions and its motives and its designs to justify nuclear cooperation of an allegedly peaceful nature."[18]

But Communist China's leadership has no hesitation in subverting American anti-proliferation goals, if the price is right. According to the CIA, "China has established itself as Iran's principal supplier of nuclear technology."[19]

In the early 1980s Beijing and Tehran signed a secret nuclear cooperation agreement. Under the agreement China has trained a number of nuclear engineers from the Atomic Energy Organization of Iran in nuclear reactor design and research.[20] Chinese technicians also began quietly helping Iran build the Esfahan Nuclear Technology Center three hundred miles south of Tehran.[21]

In January 1990 China and Iran expanded their weapons cooperation as the Iranian defense minister and the deputy director of COSTIND announced a ten-year "military technology transfer agreement."[22] Under this agreement Chinese nuclear specialists began to appear at the Qazvin Nuclear Research Center west of Tehran, and later at Karaj, another location where the Chinese supplied nuclear equipment.[23] Internationally known Iran specialist Kenneth Timmerman reports that Chinese weapons technicians have been providing further assistance to the Iranians: in uranium mining techniques and on how to turn natural uranium into uranium hexafluoride, a key ingredient in nuclear weapons-making.[24] The China National Nuclear Corporation has announced that it has been helping Iran—as well as Libya, North Korea, and Pakistan—search for uranium.[25]

The Iranians have cleverly covered up their nuclear program, thereby showing the problems inherent in international inspection. In February 1992, months after Timmerman reported that Chinese nuclear technicians were "present in force" at the Qazvin Nuclear Research Center,[26] inspectors for the International Atomic Energy Agency (IAEA) finally had a look. IAEA Deputy Director General for Safeguards Jon Jennekens told the press, "We are pleased to confirm that there doesn't seem to be a shred of evidence of any of these misleading misrepresentations." But later it was discovered that he had not even visited the suspect nuclear facilities; he had been taken to a recreational facility with a similar name! He hadn't known where he was because, over the intense objections of IAEA staff, Jennekens felt that it would indicate a "lack of trust" if he took along the handheld Global Positioning System (GPS) locators used by allied forces to find their way in the desert during the Gulf War.[27]

The PRC's clandestine nuclear aid to Iran continues, as does the struggle to uncover it. In the spring of 1998, just weeks after winning yet another Beijing pledge to halt assistance to Iran's nuclear program, allied intelligence discovered secret government-to-government negotiations designed to transfer hundreds of tons of nuclear bomb-making materials to Iran.[28] More denials, more promises—until the next time the Chinese got caught. A year later, Beijing was discovered peddling specialty steel and nuclear-grade graphite-making equipment to Tehran.[29]

Nor have the Chinese restricted themselves to providing nuclear aid. The heads of allied and Russian intelligence agencies have repeatedly warned of Iran's rapidly increasing chemical and biological warfare capabilities. According to CIA Director John Deutch, Iran began its biological warfare program in the early 1980s and now has "the capacity for large-scale BW [biological warfare] agent production." In January 1997 Secretary of State Madeleine Albright, responding to written questions from the Senate Foreign Relations Committee, stated that, as classified reports reveal, Chinese firms have transferred biological warfare equipment to Iran.[30] The CIA is particularly concerned that in the next few years Iran may develop a biological weapon warhead for its ballistic missiles.[31] Cruise missiles are another possible Iranian biological weapon delivery system, since Iran, Syria, and China are thought to have such weapons under development.[32]

With regard to chemical weapons, the CIA reports, "Iran's CW [chemical warfare] program is already among the largest in the Third World, yet, it has continued to expand and become more diversified, even since Tehran's signing of the Chemical Weapons Convention in January 1993."[33] The CIA believes Iran has the capacity to produce a thousand tons of chemical weapons per year and may have the ability to deliver them by ballistic missile.[34]

NATO's intelligence community has no doubts as to the identity of Iran's chief chemical weapons suppliers—Chinese government arms dealers. In 1993 then–CIA Director James Woolsey named the Chinese as the culprits;[35] Defense Intelligence Agency Director James Clapper seconded him in 1994.[36] Germany's BND intelligence agency reports that Chinese firms are helping Iran build a major chemical weapons facility at Parchin, twenty-five miles south of Tehran.[37]

In the fall of 1995 Representative Ben Gilman (R-New York), chairman of the House International Affairs Committee, elicited the following admission from Deputy Assistant Secretary of Defense Bruce Reidel:

> In the chemical arena, we have seen evidence that China has provided some assistance or Chinese firms have provided some assistance, both in terms of the infrastructure for building chemical plants and some of the precursors for developing agents.[38]

Four months later, the full extent of China's assistance became apparent. Quoting American intelligence officials, the *Washington Post*'s R. Jeffrey Smith reported that Chinese firms were selling Iran "*virtually complete factories* suited for making deadly poison gases [emphasis added]."[39] Smith revealed that Washington had detected a "steady flow of Chinese chemical-related equipment to Iran" and that this equipment had "been sold directly by Chinese firms to Iranian organizations affiliated with the military or the Revolutionary Guards."[40] American officials have never denied Smith's story.[41]

The PLA-associated Norinco has long been identified by U.S. officials as a major supplier of chemical weapons equipment to Iran.[42] In January 1997 Hong Kong Customs officials seized documents from a Norinco subsidiary, Rex International, showing that it had shipped to Iran high-grade seamless steel pipes that were specially treated to handle corrosive materials. U.S. officials believe the equipment "either went to an Iranian chemical weapons plant or to a missile plant."[43]

Because of these chemical weapons exports, and under intense pressure from Senator Robert Bennett (R-Utah), the Clinton administration in May 1997 was forced to place sanctions on two Chinese companies, the Nanjing Chemical Industries Group and the Jiangsu Yongli Chemical Engineering and

Technology Import/Export Group, as well as a PRC-affiliated company, Cheong Yee Limited in Hong Kong.[44]

The PRC has also helped Iran develop its missile program. As early as September 1986 Senator Frank Murkowski (R-Alaska), then-chairman of the Senate Foreign Relations Subcommittee on East Asia, publicly warned that China was in serious negotiations with Iran for the sale of ballistic missiles and related technology.[45] Chinese technicians began by helping North Korea establish a Scud-B missile production line in Iran and assisting in the development of the *Oghab* (Eagle) rocket system.[46]

In the mid-1980s Chinese missile firms began to produce their own modern "M" series ballistic missiles—M-7, M-9, M-11, and M-18. All of these missiles are "NBC capable"— that is, they can take a nuclear, biological, or chemical warhead. The Chinese M-7 missile is a highly mobile, solid-fuel, surface-to-surface, two-stage rocket system with a range of 100 miles. The M-9 and M-11 missiles are equally mobile, single-stage, solid-fuel missiles with ranges of 350 miles and 150 miles, respectively. The M-18 is a two-stage, longer-range version of the M-9.[47] Ninety M-7s were sold to Iran in 1992, and Chinese companies are assisting Iran with domestic production of the M-9, M-11, and M-18.[48] The M-18 being constructed in Iran, to be called the Tondar-68, would have a range of 650 miles, bringing Tel Aviv within easy range.[49]

The Chinese may be helping the Iranians realize their ambitions for even longer-range missiles, in the 1,000- to 2,000-mile class. In June 1995, quoting a classified CIA report entitled "China-Iran Missile Technology Cooperation: A Time-line Approach," *Defense News* revealed Chinese military companies' extensive effort to help Iran develop an indigenous ballistic missile capability.[50] Chinese ballistic missile technicians are reportedly working at a defense complex in Karaj, just outside Tehran.[51] American specialists believe Chinese arms firms have supplied the Iranian missile program with "solid fuel technology, gyroscopes and technology to build and test guidance-system components."[52] Quoting CIA sources, the *New York Times*'s Elaine Sciolino reported in June 1995 that the PRC had shipped dozens, if not hundreds, of critical computerized machine tools and missile guidance systems to Iran in the previous twelve months.[53] A top-secret NATO intelligence assessment, leaked to the Italian press, asserts that Iran will have its own 2,000-mile ballistic missile thanks to a Chinese "command and control system disguised as satellite technology."[54] According to General Peay of the U.S. Central Command, in 1996 the United States discovered that Iran was building tunnels along its coastline to hide and protect long-range ballistic missiles.[55]

What is perhaps most frightening is what we don't know about Chinese proliferation activities. Throughout 1997 the U.S. intelligence community continued to predict that Iran would not have an intermediate-range ballistic missile for ten years, but in July 1998 the Iranians lit one off.[56] The Iranian missile test-fired, known as the Shehab 3, seems to have been produced with assistance from China, North Korea, and Russia.[57] It can reach all of the Middle East, making U.S. forces in the region vulnerable to a ballistic missile attack. Much longer-range versions, the Shehab 4 and 5, are under development, which will ultimately give Iran the capability of hitting the United States itself.[58]

For more than a decade the United States has been concerned about Poly Group's sales of advanced conventional weapons to Iran, particularly cruise missiles like the Silkworm—a thousand pounds of high explosive, cruising at five hundred miles per hour, and homing in on the target with radar and infrared targeting devices—and its successors.[59] In the fall of 1987 Iranian Silkworms hit two oil tankers—one American-owned and the other flying the American flag—in the Persian Gulf. Over the course of the 1980s the Chinese exported in excess of $1 billion worth of Silkworms to Iran.[60]

The Silkworm has effectively been succeeded by the Chinese Eagle Strike missile, a much more dangerous weapon modeled on the French Exocet. The Eagle Strike comes in two versions—a solid-fuel, rocket-powered model (designated C-801 by NATO) and a turbojet-powered model (C-802). The two missiles are virtually identical, though the turbojet of the C-802 gives it a longer range—eighty miles—than its rocket-powered sibling. These missiles are much smaller than the Silkworm and are consequently more difficult to pick up on radar. When it acquires the target, the C-802 is programmed to dive down to just fifteen feet off the water, making it very difficult to defend against.

> With every Chinese arms deal the United States and its allies are in more and more danger.

The C-802s are an example of the problems inherent in estimating Chinese capabilities. At the end of 1993 the Office of Naval Intelligence estimated that the C-802 would be operational in 1998,[61] but in fact the missiles showed up on board Chinese-built Iranian patrol boats in the summer of 1995.[62] With fifteen thousand American sailors and airmen in the Gulf to worry about, Admiral Scott Redd, then–commander-in-chief of the U.S. Fifth Fleet, held three separate news conferences in 1996 to warn of the C-802s and a rocket-propelled sea mine the Chinese had also delivered to Iran.[63] CIA Director John Deutch pointed to the dangerous Chinese cruise missile sale in testimony before the Senate,[64] and Undersecretary of State Lynn Davis told the House of the "evidence" of the sale.[65] The State

Department has disclosed to Congress that the China National Precision Machinery Import-Export Corporation is the "logical originator" of the C-802s in Iran.[66]

With the C-801 and C-802 now in operation, the Iranians pose a direct threat to American armed forces. In private briefings on Capitol Hill, Office of Naval Intelligence officials indicated that land-based batteries of C-802s would give the Iranians a weapon of "greater range, reliability, accuracy and mobility" than anything currently in their inventory.[67] When, in June 1997, the Iranians test-fired a C-801 from a jet aircraft,[68] a senior U.S. military officer commented, "You now have a 360 degree threat," referring to Chinese cruise missiles that could be fired at U.S. forces from the land, sea, and air.[69]

The danger to American forces and Gulf shipping extends beyond the Eagle Strike missiles. China is developing a third generation of cruise missiles, the C-101 and C-301 series, that are ram-jet powered and supersonic, traveling at speeds in excess of Mach 2. The C-101 has a range of thirty miles, the C-301, seventy-five miles—plenty long enough for the narrow waters of the Persian Gulf.[70] These missiles have armor-piercing warheads twice the size of the C-802s, and their ability to dive down to low altitude was deliberately chosen to make them more difficult to defend against.[71] A supersonic sea-skimming missile allows virtually no response time—a potential nightmare for American naval planners in the Gulf.

The nightmare does not end there. According to a highly classified Pentagon study, American intelligence believes that China, Iran, and Syria are aggressively engaged in a cooperative program to develop an even more sophisticated cruise missile. This model, expected to become operational between the years 2000 and 2010, would have some form of Russian-derived stealth technology,[72] have a range of 1,500 miles, and

صاروخ C101 مضاد للسفن فوق الصوت

This image of the C-101 missile comes from a brochure put out by a PLA-associated arms company.

carry a nuclear, chemical, or biological warhead. The propulsion system may be based on a new Chinese mini-turbojet engine now in the prototype stage.[73] The Pentagon believes no existing anti-missile system could halt a barrage of such weapons.[74] China has already provided Iran with cruise missile navigation components (including GPS technology), information on propulsion techniques, and production equipment.[75]

In September 1996 the Iranians and the Chinese announced a $4.5 billion arms deal that specifically included missile systems.[76] From what we know of the Chinese and the Iranians, it appears the problem is only getting worse—and with every transaction the United States and its allies are in more and more danger.

LIBYA: BACKING QADDAFI
Always eager for trading partners, Chinese arms dealers are right in the middle of Libya's arms buildup.[77]

Libyan leader Muammar Qaddafi is nothing if not auda-
cious. During a trip to Beijing in 1982 he offered to buy two
nuclear weapons for a billion dollars in cash. On another occasion he tried to obtain Chinese CSS-2 intermediate-range ballistic missiles. Both times he was turned down.[78]

Although Libya used to be a Soviet client-state, the new
leadership in Moscow has decided to live up to the United
Nations arms embargo imposed after
Libyan terrorists bombed Pan Am Flight **Chinese missile transfers**
103 in 1988. But the PRC respects noth- **could enable Libyan leader**
ing, including the United Nations, when **Muammar Qaddafi to fol-**
it comes to making money. American **low through on his threats**
satellites caught a Chinese ship brazenly **to terrorize all of Western**
unloading arms at a Libyan port several **Europe.**
days after the embargo went into effect.[79]

Chinese assistance to Qaddafi's chemical weapons ambitions
seems the most worrying. In 1989, after serious international
criticism led German authorities to terminate their companies'
assistance to Qaddafi's giant poison gas factory at Rabta, Chinese companies immediately moved in to fill the void. In 1990
American intelligence reported that Chinese arms companies
had been supplying more than a dozen critical chemicals for
Rabta.[80] And Chinese specialists are also reportedly working
with Libya on chemical weapons programs.[81] At a March 1990
Senate Foreign Relations Committee hearing, Senator Joseph
Biden (D-Delaware) pressed the State Department on the issue,
but State reported that the Chinese had no response to American
complaints.[82]

Seven years later, nothing had changed. On her first trip to
Beijing as secretary of state, Madeleine Albright raised the issue
of continued Chinese arms sales to Libya but received no
response.[83] Two months later, she repeated her concerns, this

time in Washington to Foreign Minister Qian Qichen, but again to no avail.[84]

With the factory at Rabta under the international micro-scope, Qaddafi and the Chinese have moved their operations to at least two new underground facilities. In mid-1990 the Bush administration told the press that Chinese state-owned companies were in the process of delivering ten thousand tons of chemical weapons precursors to a new poison gas plant Qaddafi was building at Sabha, on the outskirts of Tripoli, the Libyan capital.[85] In 1996 German intelligence reported that Chinese companies are providing components and expert help to Qaddafi's massive new poison gas plant under a mountain at Tarhunah, fifty miles south of Tripoli.[86]

Unsurprisingly, the Chinese have sold the Libyans lithium hydride—the same useful chemical they sent to Iraq—benefit-ting Qaddafi's nerve gas, missile, and nuclear programs.[87]

The Chinese could well be supplying Qaddafi with missiles, also. Libya's Soviet-made Scud missiles have a range of only eighty-five miles—as the Libyans discovered when their missile strike on an American military base on the Italian island of Lampedusa fell short.[88] If Qaddafi wants to cause the mass destruction of European cities, as he has threatened,[89] he will need to gain a longer-range punch—at least 750 miles.

In fact, Qaddafi could already have the long-range capabili-ties he needs. In 1989 he reportedly put up $170 million for 140 Chinese M-9 missiles, 60 of which were to be delivered to Libya and 80 to Syria.[90] Both the Bush administration and the Israeli government complained about these missile deals at the Beijing regime's highest levels but received the usual Chinese bluster. Moreover, it has been reported that the PRC has contracts to provide missile technology to Libya,[91] and Chinese specialists

Dear Reader:

Thank you for purchasing this Regnery book. Since 1947, we have published books on a wide variety of subjects. Often on the cutting edge of American and global affairs, our books are known for challenging the status quo. The book you have purchased is no exception.

If you would like to know more about our books, please fill out this postcard and drop it in the mail. Thank you.

Sincerely,

Alfred S. Regnery

Alfred S. Regnery
President & Publisher

REGNERY
PUBLISHING, INC.
Established 1947

Name ☐ Mr. ☐ Ms. ☐ Mrs. _____

Address _____

City _____ State _____ Zip _____

E-Mail _____

I received this card in the book titled _____

were spotted conducting research and development work in Libya in the late spring of 1998.[92]

In 1994 British Defense Secretary Malcolm Rifkind told the Jewish Board of Deputies in Glasgow that Libya has missiles capable of reaching Spain, southern France, and Italy—far beyond the range of Scud-Bs.[93] Rifkind later told the British Parliament that Libyan missiles could reach Gibraltar, nine hundred miles from Tripoli.[94]

But where did Libya get these missiles? The secretary didn't say, but the answer, it seems, is the PRC. Quoting Western intelligence, the *Times* of London reported in 1995 that Libya and Iran were suspected of cooperating in the development of long-range ballistic missiles, and that Iran was willing to provide Chinese and North Korean missile technology to Tripoli for the sum of $31 million.[95] Such a deal would benefit all four parties—Iran, Libya, North Korea, and the PRC. Libya would get the missiles it desired; Iran would get the cash it needed to pay the Chinese and North Koreans to help develop its own missile system; the PRC would receive yet another lucrative arms contract; and North Korea would also receive much-needed cash.

Perhaps based on NATO intelligence, a Spanish newspaper has predicted that by 2006 Libya is expected to have missiles that can be armed with weapons of mass destruction and that have a range of 1,900 miles.[96] In other words, Qaddafi could terrorize all of Western Europe. The Chinese are helping him follow through on his threats.

MISSILE BUILDUP IN SYRIA

Because Syria shares a border with Israel, the United States is particularly sensitive to any sophisticated arms sales to Syrian dictator Hafez al-Assad's military machine. With regard to nuclear issues, U.S. policy toward Syria is the same as the policy toward Iran: zero tolerance. And China's policy toward Syria is the same as its policy toward Iran: whatever the traffic will bear. Over America's strenuous objections, the Chinese kick-started Syria's nuclear program in 1991 with a small, thirty-kilowatt nuclear reactor. In short, China has initiated the nuclear science program of another terrorist regime. Syria did sign a safeguards agreement with the IAEA,[97] but it is far too risky to allow the Syrian regime a nuclear program because of the difficulties involved in international monitoring and inspection.

Most of the controversy surrounding Chinese illicit arms sales to Syria is missile-related. As early as the summer of 1988 the Reagan administration was reporting that a major Poly Group–brokered China-Syria missile deal was imminent.[98] One exasperated Reagan official told the press, "The Chinese seem to be driven blindly by moneymaking."[99] A year later well-respected Arab newspapers reported that China and Syria had signed a major deal for Syria to receive the M-9 missile.[100] From 1989 to 1992 the Bush administration and the Israelis tried to induce Beijing and Damascus to cancel the arrangement,[101] and as late as October 1991 Secretary of State James Baker's trip to Beijing was being held up because of the pending PRC-Syria missile deal.[102]

There the trail goes cold. If Beijing actually transferred M-9 missiles to Syria, no credible public record of it exists. Japanese intelligence reports that Chinese Premier Li Peng promised the Syrians sophisticated surface-to-surface missiles during a visit to Damascus in early July 1991.[103] In September 1991 the head of German intelligence told reporters that the Chinese had transferred critical missile equipment called Transporter Erector

Launchers (TELs), but not missiles, to Syria.[104] In the spring of 1992 William Safire of the *New York Times* revealed that Chinese experts were helping construct two underground missile plants in Syria.[105] According to Safire, one week after pledging to Secretary Baker that they would not transfer M-9s to Syria, "the Chinese secretly agreed to help the Syrians construct their own missiles locally."[106] The Beijing Wanyuan Industry Corporation even sent the Syrians a financial specialist to show how to keep a missile development program under budget![107]

Israeli specialists believe that Syrian ballistic missiles may have Chinese chemical or biological warheads.[108] In early 1992 then–CIA Director Robert Gates testified before Congress that Syria was seeking assistance "for an improved capability with chemical and biological warheads" from Chinese firms.[109]

A trickle of public accounts of missile parts making their way from China to Syria supports the contention that the Chinese have helped the Syrians' missile program. After agreeing to abide by the terms of the Missile Technology Control Regime (MTCR), Beijing was observed by American and Japanese intelligence shipping thirty tons of a prohibited rocket fuel, ammonium perchlorate, to Damascus.[110] More recently, in the summer of 1996 Bill Gertz of the *Washington Times* reported that the Chinese had shipped missile guidance equipment to Syria, which set off a frantic spin control operation by the Clinton administration.[111]

Now we know that the Chinese and Syrians are circumventing proliferation restrictions through third-party transfers. In August 1999 ABC News reported that, according to U.S. and Israeli intelligence, "Syria is obtaining Chinese medium-range, mobile-launch missile technology through a circuitous route that involves Iran, Pakistan, and North Korea"—all major customers of PLA arms dealers.[112]

BROKEN PROMISES

In January 1993 the Arms Control and Disarmament Agency
(ACDA) issued a report stating that the PRC had an offensive
germ warfare program that had for years violated its commitments under the 1972 Biological Weapons Convention (BWC).
Ambassador Bradley Gordon, the ACDA's assistant director
for nonproliferation, pushed the report through the system and
got President George Bush to sign it on his last day in the White
House.[113]

Ever since, ACDA reports to Congress have identified the
PRC as a violator of its germ warfare commitments. An ACDA
report released in the summer of 1997 contains the following
"finding":

> The United States believes that China maintained an offensive
> BW [biological warfare] program prior to 1984 when it became
> a party to the BWC, and maintained an offensive BW program
> [that] included the development, production and stockpiling or
> other acquisition or maintenance of biological warfare agents.
> China's... declarations have not resolved U.S. concerns that
> China probably maintains its offensive program. The United
> States, therefore, believes that in the years after its accession to
> the BWC, China was not in compliance with its BWC obliga-
> tions and that it is highly probable that it remains noncompli-
> ant with these obligations.[114]

This is bureaucratese for "They cheated and they lied."

In fact, the PRC has an extensive illegal germ warfare program. According to the U.S. Department of Defense, this
includes "manufacturing infectious micro-organisms and tox-
ins."[115] The Defense Department adds, "China has a wide range
of delivery means available, including ballistic and cruise mis-

siles and aircraft, and is continuing to develop systems with upgraded capabilities."[116] According to the 1999 book *Biohazard* by Ken Alibek and Stephen Handleman, China is still producing hemorrhagic viruses at a secret laboratory in Central Asia. China, like the Soviet Union, has reportedly had a series of accidents at its biological warfare facilities that has killed hundreds of people.[117]

It is bad enough that the PRC has a domestic germ warfare program that violates its international commitments, but, as Secretary of State Madeleine Albright confided to the Senate, the United States knows that in at least one circumstance Chinese companies have secretly exported germ warfare equipment to a terrorist country, Iran.

Why didn't Vice President Gore call for sanctions against the Chinese and defend the integrity of legislation he sponsored as a senator?

Worse still, the Chinese government has also violated its international nuclear nonproliferation obligations, as dictated by the Nuclear Nonproliferation Treaty, which the PRC signed in 1992.[118] Even the U.S. State Department has admitted this in testimony before Congress. In 1996 Undersecretary of State Lynn Davis explicitly confirmed to the House International Relations Committee that China is in violation of its international commitments under the Nonproliferation Treaty because it sent Pakistan—a "nonnuclear weapons state," under the terms of the treaty—five thousand samarium-cobalt ring magnets, which are critical to producing bomb-grade enriched uranium.[119]

The 1995 ACDA report states, "Since China's accession to the NPT [Nonproliferation Treaty], it appears that China may have continued to assist Pakistan's unsafeguarded nuclear program and may have continued contacts with elements associated with Pakistan's nuclear weapons related program."[120]

The PRC has similarly violated its pledges in regard to its chemical weapons program. In 1993 Communist China signed the Chemical Weapons Convention, a treaty that prohibits transferring "chemical weapons to anyone," whether "directly or indirectly."[121] But as we have seen, *after* the PRC signed the Chemical Weapons Convention, PLA-associated companies shipped "virtually complete" poison gas factories to Iran.

The PRC did not immediately ratify the Chemical Weapons Convention upon signing it; still, Beijing's chemical weapons transfer to Iran after it had signed the treaty violates international law. Under Article 18 of the Vienna Convention on the Law of Treaties, any country signing a treaty has an "obligation not to defeat the object and purpose of a treaty prior to its entry into force."[122] Certainly, exporting a "virtually complete" nerve gas plant to a terrorist country such as Iran would defeat the purpose of the Chemical Weapons Convention.

The Chinese record is no better when it comes to missile proliferation. The Missile Technology Control Regime (MTCR) commits members not to export ballistic missiles or related technology with certain capabilities in range (three hundred kilometers) or throw weight (five hundred kilograms), but the M-11 missiles the Chinese have transferred exceed those parameters. Although the MTCR is officially a voluntary arrangement rather than a treaty or international agreement,[123] the Chinese government has violated the written promises it gave the United States in February 1992 and again in October 1994, which create valid bilateral agreements under international law.

In short, in all the major categories of international arms control agreements—biological weapons, nuclear weapons, chemical weapons, and missiles—the Chinese government has violated its most solemn pledges. And the PRC has given no indication it intends to fulfill any of its promises in the future.

THE RECORD

Under American law, it is illegal for a foreign arms broker to sell chemical weapons, biological weapons, nuclear weapons, ballistic missiles, or any of the related technology,[124] and it is further illegal to sell advanced conventional weapons, including cruise missiles, to Iran or Iraq.[125]

Consider the following:

- President Clinton, his secretary of defense, his secretary of state,[126] and a commission he appointed[127] have all identified the proliferation of weapons of mass destruction as a very real threat to the life of every American citizen, our military forces, and our allies abroad.
- During the time President Clinton and Vice President Gore have been in office, scores of credible reports have revealed that Chinese arms firms are delivering weapons of mass destruction and missiles to terrorist regimes, in flagrant violation of American law and international commitments.[128]
- The Clinton-Gore administration has imposed sanctions just once on Chinese arms brokers for missile deliveries to Pakistan and once for chemical weapons deliveries to Iran;[129] it has imposed no sanctions for germ warfare equipment sales or for cruise missile transfers to Iran.

In the case of the cruise missile sales to Iran, there is no question that the missiles (C-801s and C-802s) have been shipped to Iran or that they pose a direct threat to the fifteen thousand American troops serving in the Persian Gulf region. Sanctions could clearly be justified on the basis of the Gore-McCain Act—named after its principal drafters, then-Senator Al Gore (D-Tennessee) and Senator John McCain (R-Arizona).

Hundreds of these missiles have been sold to Iran on Bill Clinton and Al Gore's watch. Why didn't the vice president speak up to defend the integrity of his own legislation, not to mention the safety of the American service men and women in range of these weapons?

There may be an explanation: Democratic National Committee fund-raiser Johnny Chung told the federal grand jury that hundreds of thousands of dollars in illegal campaign contributions for the 1996 Clinton-Gore reelection effort came in from Chinese military intelligence. PLA spy Lieutenant Colonel Liu Chaoying oversaw this operation; she was also a major arms dealer for China National Precision Machinery Import-Export Company, the manufacturer of the cruise missiles. Poly Group brokered the missile deal with Iran, and Poly's chairman, Wang Jun, appeared at a White House coffee fund-raiser in February 1996. That happened to be the same time as the U.S. Navy was beginning to brief Congress on the threat posed by C-802 missiles.[130]

Was it all a coincidence? Doubtful, in our view.

PART III
THE TARGETS

CHAPTER 8
TARGETING AMERICA:
THE REVOLUTION IN MILITARY AFFAIRS

Information warfare and electronic warfare are of key importance, while fighting on the ground can only exploit the victory.
Hence, China is more convinced [than ever] that as far as the
PLA is concerned, a military revolution with information warfare as the core has reached the stage where efforts must be
made to catch up with and overtake rivals.[1]

—General Liu Huaqing, vice chairman of the
Central Military Commission, 1995

On Wednesday night, June 16, 1999, officials at a
California sanitation plant decided to check its
computers for Year 2000 (Y2K) compliance. They
were testing a back-up electrical system when they received a
frantic midnight call from a park ranger: Raw, untreated sewage
was pouring out of a manhole cover and spilling into a park.
Later it was estimated that four million gallons of sewage had
been released. A computer had mistakenly closed a gate that
should have remained open to control the transfer of sewage. A
programmer's error fifteen years earlier seems to have been the
culprit. It cost taxpayers about $100,000 to clean up the mess.[2]

This is just one of a number of unfortunate accidents that have occurred as businesses and governments test for Y2K compliance. But Y2K accidents are just that—accidents. Suppose, however, that a competent and motivated hostile force was able to manipulate modern computer systems. Such a force could conceivably do the following:

- Change the dose levels in prescription medicines at pharmaceutical plants so that thousands of people would be poisoned[3]
- Infiltrate the manufacturing process for baby food so that the standard components would be increased by 400 percent—to toxic levels[4]
- Taint the processed-food industry for restaurants, hotels, hospitals, and retirement homes
- Subtly change airport radar signals so that air traffic controllers would unknowingly put passenger planes on the same flight path
- Open the electronic gates and fences at a number of jails and prisons around the country simultaneously, overwhelming law enforcement officials[5]
- Stage a surprise attack on many of the automated gasoline refineries in the nation, causing enormous, out-of-control fires that would inundate emergency officials and lead to immediate gasoline rationing[6]
- Contaminate the city water systems, turn the valves backwards at the sewer systems, shut down the electric power grid, and overload the natural gas pipeline system
- Loot bank accounts, transferring all funds overseas[7]
- Attack individuals' identities, eliminating their Social Security records, Veterans Department records, driver's license numbers, bank accounts and credit card numbers, and so on[8]

All of these are examples of what is known as "information warfare," or more specifically "offensive information warfare."

Is there something to this, or is it just the product of an overactive imagination?

The concept of information warfare—and in particular, offensive information warfare—is perhaps America's most highly guarded military secret today.[9] Most experts in the field believe that the United States is currently the world's information warfare leader.[10] But the interconnected nature of modern American society makes the United States "the most vulnerable country in the world" for this sort of warfare, according to the former director of the National Security Agency, our premier electronic warfare agency.[11] *Every one* of the examples above has either been tried successfully or will be within the capability of hostile forces in a short period of time, and every one is of deep concern to the American government.[12] The problem is real.

The Chinese People's Liberation Army has the world's largest information warfare program, after the United States.

More alarming, the Chinese People's Liberation Army has the world's largest information warfare program, after the United States.[13]

WARFARE FOR THE TWENTY-FIRST CENTURY

Andy Marshall is the director of the Office of Net Assessment, a tiny Pentagon office that looks to the future of warfare and makes certain the United States is prepared to meet any challenge.[14] In recent years Marshall has revealed something called the "Revolution in Military Affairs," or RMA for short. In 1995 he told the Senate that over the next twenty to fifty years there will be a revolution in the way wars are fought; in essence, warfare will be conducted on a high-technology battlefield such as we've never known before.

Two principal elements of the RMA will be "long-range precision strike weapons" and "information warfare." Today, long-range precision strike weapons include bombs and missiles, many of them laser guided with the aid of satellites or orbiting aircraft. Laser-guided bombs have been in the U.S. inventory for more than thirty years, but in time we expect that American and foreign military forces will field more exotic weapons—ones that fire a directed energy burst, for example, rather than any sort of bullet, shell, or rocket.

Twenty-first–century weapons will be linked to very effective sensors and command and control systems. "Sensor" may ultimately mean an unmanned aerial vehicle the size of a large flying insect that can transmit television pictures, sound, and even the smell of whatever is in its range. Within ten years the individual American soldier will be "digitalized." That is, he or she will be in constant contact with commanders through miniature TV cameras that allow commanders to see what the soldier sees, day or night. Chinese military thinkers envision a day in the not-too-distant future when a soldier's uniform will be heated or cooled like a space suit and the soldier will have an individual flight capability to jump over obstacles.[15]

But for every communication—from the sensors to the command post, or between the command post and the soldiers in the field—there is an opportunity for a hostile power to intercept it. Once intercepted, the transmission can be manipulated or jammed. The key to this warfare will be to protect your own systems and penetrate the enemy's systems. Alvin and Heidi Toffler, noted commentators on the future of warfare, point out that "any military... has to acquire, process, distribute, and protect information, while selectively denying or distributing it to its adversaries and/or allies."[16] Information warfare will be used to jam, corrupt, or manipulate the enemy's command and control system as well as attack the opponent's civilian economy and government services.

There is more to information warfare, however, than just jamming, cutting down telephone lines, and shutting down the electric power system. As U.S. Army Field Manual 100-6 points out,[17] during the Civil War, Confederate General J. E. B. Stuart's cavalry forces often seized Union telegraph offices but kept the telegraph working in order to transmit false orders and scramble the Union Army's command and control system. General Stuart's cavalry was practicing deception, electronic warfare, psychological warfare, and special operations,

> **The American economic, political, and social system is essentially unprotected against a Chinese information warfare attack.**

all part of modern American strategy on information warfare.[18] Today such warfare also includes "computer network attack," the precise parameters of which are highly classified.[19]

The PLA uses the same strategy of deception, electronic warfare, psychological warfare, and special operations: Once inside an enemy's command and control network, the intruder could give enemy troops false orders to fire on each other or march into an ambush. Rather than jamming a radar system, the intruder could make the operator think a "virtual military unit" or "virtual fleet" is attacking from the south when it is actually moving in from the north. A leading PLA authority has written that it may even be possible to have the image of an opposing supreme commander appear on television ordering his troops to surrender.[20]

Frightening as some of these possibilities may be, in our view the far greater risk to the United States involves the American economic, political, and social system, which is essentially unprotected. As the PLA notes, "America's economic system is extremely vulnerable to information attacks."[21] The PLA even envisions imposing an "economic information blockade," noting that the more a country depends on imports "the greater the damage to the economy."[22] The United States, a nation that imports well over 50 percent of its oil and gas, could be quite vulnerable.

In fact, it is our judgment that information warfare from the People's Republic of China is an unheralded national security threat to the United States and the rest of the democratic countries.

THE PLA AND INFORMATION WARFARE

In a report to Congress in 1998, the Department of Defense said:

> In recent years, the PLA has shown an exceptional interest in information warfare (IW) and has begun programs to develop IW capabilities at the strategic, operational and tactical levels as part of its overall military modernization effort.[23]

Our military academics concur. American PLA specialist Dr. Michael Pillsbury puts China right behind the United States in information warfare capability; Dr. James Mulvenon of the RAND Corporation says that the PLA is third after the United States and Russia, pointing out that the PLA "has an active offensive IW program and has devoted significant resources to the study of IW."[24] The director of the Army War College's Strategic Studies Institute, Colonel Larry Wortzel, agrees that China has an active information warfare program, noting, "Indeed, the PLA is using the armed forces of the United States, their doctrine and strategic orientation as its model for what constitutes a power-projection force."[25]

Chinese military writings are filled with articles describing information warfare, its possibilities as an offensive weapon, and the dangers it poses to the defense. *Liberation Army Daily*, the PLA's leading newspaper, has run dozens of major articles in the past five years on information warfare. Leading Chinese defense scientists have also endorsed information warfare as the wave of the future,[26] and, as we have seen from General Liu Huaqing's emphatic support for information warfare, the top PLA brass is behind the drive to gain the edge in the Revolution in Military Affairs. In fact, at least one unclassified PLA article claims that the Chinese are already conducting information warfare military exercises.[27]

Although information warfare represents the most modern warfighting capabilities, it also fits within China's ancient military tradition. Sun Tzu wrote his essays on "The Art of War" about 2,400 years ago. Two of his famous maxims—"All warfare is deception" and "To subdue the enemy without fighting is the acme of skill"—apply directly to information warfare.[28] Moreover, information warfare is both less expensive than any other method of strategic strike and less time-consuming to prepare.

Of course, in order for the PLA's information warfare program to be a threat to the United States, it has to have five capabilities:

- Signals intelligence (SIGINT), which allows the PLA to intercept or otherwise break into a computer network
- Supercomputers
- Trained technical personnel
- Sophisticated weaponry
- Strategy and planning

Unfortunately, the PLA is achieving these five capabilities.

SIGINT

According to Professor Desmond Ball, writing in *Jane's Intelligence Review*, "China maintains by far the most extensive signals intelligence (SIGINT) capabilities of all the countries in the Asia/Pacific region."[29] The PLA has SIGINT sites all around China's periphery—in China's northwest region (built with American help during the Carter administration), in the South China Sea, on a Burmese island in the Indian Ocean, in Laos, and on the island of Tarawa in the Pacific. As with the United States and Russia, the PLA operates SIGINT facilities on navy ships and submarines as well as air force planes. Its efforts to gain a toehold in space have been largely unsuccessful, at least to this point.[30]

For a number of years American PLA-watchers have noticed a surprising number of high-level military delegations traveling between China and Cuba. Initial speculation centered on Cuba's expertise in biological warfare, but during his February 1999 visit to Havana, Chinese Defense Minister Chi Haotian signed an agreement for not one but two major SIGINT facilities on the island, just off American shores. One facility is apparently designed to intercept signals from American military satellites, and the other is designed to intercept American telephone calls on the East Coast. The equipment the PLA is using is so powerful that it is interfering with air traffic control as far away as New York.[31]

Without a space-based SIGINT capability, the PLA would logically try to get as close to our shores as possible. That might also explain the desperate effort by the China Ocean Shipping Company (COSCO) to acquire space at the old Long Beach, California, naval base. COSCO would naturally have some radio antennas at Long Beach to talk to its ships at sea—and to listen into voice and data transmissions among major defense contractors in the Los Angeles area.[32] The Cuba operations as

President Bill Clinton holds a White House meeting with People's Liberation Army General Zhang Wannian, whose 15th Airborne paratroopers mowed down Chinese civilians during the June 1989 Tiananmen Square massacre. This photo ran in Chinese newspapers, helping rehabilitate the image of the "butchers of Beijing."

Vice President Al Gore and then–Chinese Premier Li Peng toast an economic partnership between the United States and the People's Republic of China.

THE TIANANMEN SQUARE MASSACRE, BEIJING, JUNE 4, 1989

The thirty-foot-tall "Goddess of Democracy" was the symbol of the Chinese students' nonviolent demonstrations in the spring of 1989.

On the floor of Beijing Fuxing Hospital lie the bodies of four Chinese civilians murdered during the June 4 atrocities.

People's Liberation Army tanks rolled over Chinese civilians.

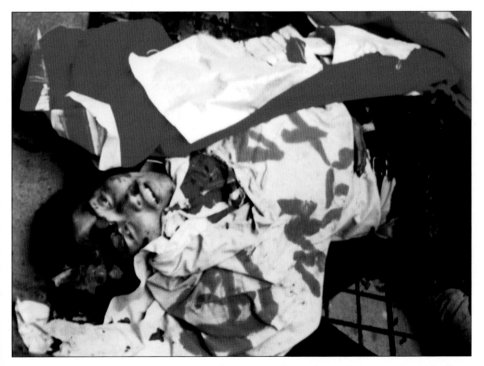

A young man lies dead in the street, the victim of People's Liberation Army bullets.

This map shows the Asian region that the People's Republic of China has sought to dominate.

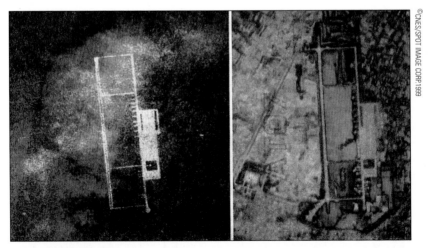

The Chinese have built a full-scale replica of Taiwan's Chingchuankang (CCK) military airfield, the island's largest air base. People's Liberation Army forces have reportedly been practicing assaults on the mock-up. At right is a satellite image of CCK; at left is the satellite photo that revealed the Chinese mock-up.

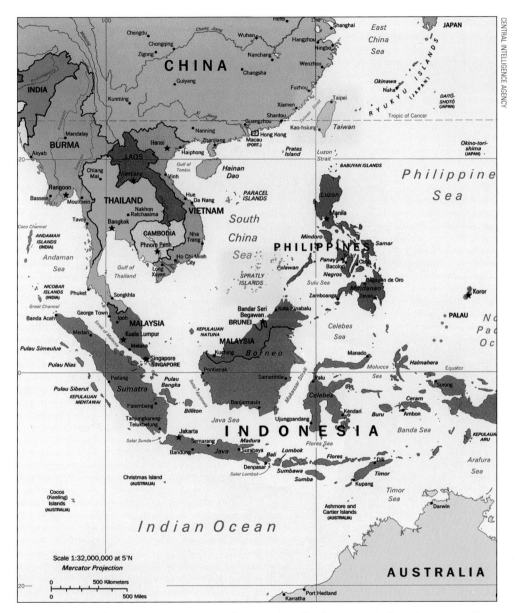

On this map of Southeast Asia, note the Spratly Islands' strategic location, as well as Taiwan's proximity to the Chinese mainland.

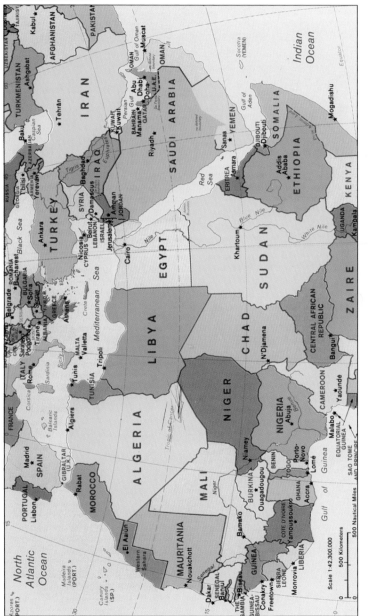

This map features North Africa and the Middle East, where a number of terrorist regimes have been prime customers of Chinese arms dealers. Thanks to Chinese weapons proliferation, the United States and its democratic allies are vulnerable to attack.

People's Liberation Army General Chi Haotian (left) and U.S. Secretary of Defense William Cohen (right) celebrate a toast; Chinese General Xiong Guangkai (behind Cohen, partially obscured) looks on. The hard-line Chi and Xiong both played pivotal roles in the brutal Tiananmen Square massacre.

President Clinton and Chinese President Jiang Zemin review People's Liberation Army troops in Beijing.

well as COSCO's operations on the West Coast, in Panama, and in the Caribbean may well fit a pattern of interlocking SIGINT operations in our part of the world.

SIX HUNDRED SUPERCOMPUTERS

The House of Representatives' bipartisan Cox-Dicks Committee, named to investigate "U.S. National Security and Military/ Commercial Concerns with the People's Republic of China," devoted an entire chapter of its report to the question of super-computer sales to China. The committee noted that virtually no domestic supercomputer industry exists in the PRC today,[33] and that the PRC had almost no supercomputers as of 1996. But, the committee noted, the Clinton administration changed the export control regulations in January 1996—at the beginning of the 1996 presidential campaign—and by the end of 1998 the PRC had more than six hundred American supercomputers.[34]

In the Cox Committee's list of the military purposes for which these supercomputers could be used, nuclear weapons design received all the publicity, but information warfare led its list of troublesome applications.[35] The committee stated that supercomputers "could prove valuable" to PLA information warfare ambitions "by exploring U.S. information networks and their vulnerabilities, and the technologies that are associated with information warfare such as jammers, microwave weapons, and anti-satellite weapons."[36]

It's useful to note the Cox Committee's use of the phrase "could prove valuable." The fact is, we haven't a clue where many of the supercomputers are or how they are being used. Nuclear weapons design? Missile design? Information warfare? Intelligence collection? Designing submarine nuclear reactors? Germ warfare? The Cox Committee lists all these as possibilities, but in truth nobody knows.

And nobody knows because the Chinese government will not allow American officials to check on them.

The case involving Sun Microsystems provides some insight into how the PLA is using the computers. In late 1996 Sun sold one of its supercomputers to China through a Hong Kong reseller, and in the spring of 1997 it ended up at a major PLA-associated weapons lab in central China. American officials became suspicious and contacted the Chinese government, which promptly denied that the lab was affiliated with the PLA. But when pressure continued, the Chinese government reversed course, admitting to the Americans that it was a PLA lab and stating that Sun had known it all along. After the story hit the newspapers, this particular supercomputer was returned to the United States, and Sun (or its Hong Kong agent) paid a small fine for violating American export control regulations.[37]

> "I am aware of the fact that the West remains our chief enemy."
>
> —Jiang Zemin, president of the People's Republic of China

THE RETURNED STUDENTS

The Cox Committee estimated that "at any given time there are over 100,000 PRC nationals who are either attending U.S. universities or have remained in the United States after graduating from a U.S. university."[38] These students are "a ready target for PRC intelligence officers and PRC Government-controlled organizations, both while they are in the United States and when they return to the PRC."[39]

Usually PLA representatives are pretty careful about what they say, but sometimes they slip up. In the fall of 1994 a PLA official bragged to a Chinese reporter that the army was recruiting "returned students" who had experience in "supercomputers and artificial intelligence."[40] Perhaps the official thought it wouldn't be noticed, since he was speaking to a domestic Chinese newspaper, rather than an international service, and since the story appeared in Chinese, not English. Still, it was an alarming admission that raised even more disturbing questions. If the PLA had no access to supercomputers at that time, why would it go to the expense of recruiting returned students with these skills? Did the PLA know, some fifteen months before President Clinton signed the order allowing supercomputers to be exported to China, that the Clinton administration would lift the ban?

HIGH-POWERED MICROWAVE WEAPONS

High-powered microwave weapons (sometimes known as radio frequency weapons), which the PLA calls the "superstars" of warfare, represent the new armaments that may define twenty-first–century warfare. These very dangerous weapons can jam electronic equipment by emitting an extremely powerful pulse of electromagnetic energy over a wide area, or their energy can be focused in a narrow beam for use against American satellites.[41] In 1995 a PLA weapons specialist proclaimed them "a special kind of information-intensified weapon for waging information warfare."[42]

The Congressional Joint Economic Committee held hearings in 1997 and 1998 on "Radio Frequency Weapons and Proliferation: Potential Impact on the Economy," during which five experts warned about the threat these weapons pose to the United States, particularly from a "determined, organized, well-funded adversary."[43]

The PLA has had a high-powered microwave weapon program for more than twenty-five years under the direction of a returned student who received his degree from the University of California at Berkeley in 1950.[44] In addition to its own work,[45] the PLA has access to American research—through espionage.[46] As the Cox Committee noted, the six hundred American supercomputers can be applied directly to microwave weapons to conduct information warfare. The American high-powered microwave weapon research and development Chinese spies obtained in the late 1990s and the six hundred supercomputers the PRC acquired once President Clinton lifted the export controls in 1996 give China an enormous advantage in this field. Worse, the PLA may have bought high-powered microwave "suitcase bombs" from the Russians.[47]

A PLAN OF ACTION

Any PLA information warfare plan would be a military secret of the highest order, but several open-source indicators point toward a comprehensive plan of action:

- Information warfare has the support of the top PLA brass.
- The PLA's best strategists and defense scientists have had extensive open discussions about information warfare.
- The PLA is conducting military exercises in information warfare.
- It is expanding—in Cuba—its already strong SIGINT capability.
- The PLA is buying the hardware; supercomputers aren't cheap.
- The Chinese are recruiting scientists and technicians.
- The PLA is building related weapons (high-powered microwave weapons).

Under these circumstances, it would be absurd to conclude that the PLA does not have a plan or is not close to completing one.

THE THREAT TO AMERICA

In the winter of 1994–1995, PRC President Jiang Zemin told PLA generals, "I am aware of the fact that the West remains our chief enemy."[48]

Why would the West be the "chief enemy" of the PRC? The "West" in this instance probably means the United States and its democratic allies, including the NATO countries, Australia, and Japan. We are the chief enemy of the PRC because we stand in the way of the Chinese Communists' ambition to dominate Asia. Congressmen Cox and Dicks made exactly that point twice in their unanimous report.[49]

As the chief enemy of the Chinese Communists, what can we expect? A PLA surprise attack against our greatest vulnerability. As we have seen, a surprise attack is the PLA's favorite weapon—used against Tibet, against U.S. forces in Korea, against India, even against the PRC's former ally the Soviet Union, and against another ally, Communist Vietnam. Add to these examples the "volunteer" soldiers who poured over the Burmese border in 1968 and the surprise occupation of Philippine territory in the Spratly Islands in 1995 (which we will see in the next chapter), and the pattern is clear.

We are deeply disturbed that senior PLA officers have begun to talk among themselves about a preemptive strike using information warfare. In 1996, for example, a writer in the PLA's main newspaper pointed out that "the enemy" has "reconnaissance positioning satellites, AWACs, stealth bombers, aircraft carriers, long-range precision weapons" but that the PLA does not.[50] A surprise attack, therefore, is not only justified but "is the only way to steer the course of the war in a direction favorable to China."[51]

A surprise attack, if it should come, would be aimed at the American people—that is, the home front. Since General Lee's surrender at Appomattox in 1865, the American home front has been virtually untouched. Except for those who were in Honolulu on December 7, 1941, Americans have no conception of the devastation that countless residents of Europe, the Middle East, Africa, and Asia have known.

In 1995 the RAND Corporation, a think tank associated with the U.S. Air Force, conducted an information warfare–based war game. Its conclusion: The United States homeland is quite vulnerable to an information warfare attack because information-based infrastructures are easy strategic targets for a hostile force.

PLA information warfare planners associate their operations with psychological warfare that can be "targeted at the masses" in order to create chaos on the home front.[52] For example, the attackers might deliberately leave the telephone system intact so that rumors would spread and the 911 system would become overloaded. The electric power grid could be left up so citizens would see and hear news reports of widespread destruction. According to a Pentagon report, such attacks could cause "widespread fear throughout the civilian population."[53]

Looking at all of the possible points of attack, we believe the most vulnerable points are the SCADAs.

SCADAs

"SCADA," the acronym for "Supervisory Control And Data Acquisition," refers to the computer-controlled gates that have mushroomed through government and industry. SCADAs have three main parts—a master station, one or more remote stations, and the custom computer software that runs them. In a public water system, they would control when the pumps start and stop, how much chlorine is injected into the system, which valves are open and which are shut, and so forth. According to the President's Commission on Critical Infrastructure Protection, "Cyber [computer] vulnerabilities include the increasing reliance on SCADA systems for control of the flow and pressure of water supplies."[54] SCADAs are used everywhere, from power companies to soft-drink factories, from processed-food factories to waste-water treatment plants.

Computer control systems similar to SCADAs called "Distributed Control Systems" are routinely used in large industrial plants such as food processing, pharmaceutical, and chemical facilities. Other SCADA-related computer systems control railway and trucking operations. Millions of such computer systems quietly perform their jobs every day throughout our society. But when one wrong command is given to just one system, the consequences can be dramatic—four million gallons of raw sewage can spill out on the ground in California.

> **President Clinton allowed the central nervous system of information warfare to be exported to the People's Republic of China.**

Consider an electric power grid. It might have two thousand SCADAs hundreds of miles apart, all tied to the central computer. Communications from the central station occur by radio, and typically the radio communications links are not encrypted. The SCADAs may also be maintained by outside contractors whose personnel have not received background checks. According to the President's Commission on Critical Infrastructure Protection, the systems rely on "dial-back modems that can be bypassed,"[55] which means that a determined outside team could gain access to the system by capturing the signal or by using the telephone system and the dial-back modems. Once inside, this hostile force could override the commands from the central station with potentially disastrous results.

Pharmaceutical plants, processed-food factories, waste-water treatment plants, refineries, chemical plants, power grids, oil or gas pipeline systems—all are susceptible. There could be havoc—fires, mass accidents, poisonings—if the SCADAs cannot be protected.

THE REFINERIES

As of this writing there are about 160 refineries in the United States. Some stand alone, but others are clustered together, sometimes with chemical plants.[56] Each of these refineries is dependent on hundreds if not thousands of SCADAs. SCADAs determine gasoline blends and even if gasoline or diesel fuel will be produced on a particular day. Operators use SCADAs to open and close valves in the refinery itself and in the oil and gas pipelines systems, sometimes hundreds of miles away.

We believe that these clusters of refineries are vulnerable to mass attack and that such an attack could have the effect of a small nuclear explosion. As the President's Commission on Critical Infrastructure Protection has stated, "Large refineries (greater than 250,000 barrel capacity) in California, Texas and Louisiana would be attractive targets for physical or cyber attack."[57]

An attack would work like this: An opposing computer team would enter the system on a reconnaissance mission, called a "sniffer" operation. The team would leave no trace of having come and gone but would install a "back door"[58] to make certain it could get in again—either through SIGINT, by capturing the radio signal, or through the dial-up modem and the telephone system. The opposition force could map the refinery's entire computer system.

A special operations team would gain a walk-through of the plant under the guise of seeking employment. Because high-powered microwave weapons have difficulty penetrating a cinder-block building, the operatives would be looking for windows or other entry ways into the central command post.

The focus of the attack would be the SCADAs. First, the special operations team would manipulate the SCADAs to start cascading fires and explosions. Then it would set off a high-powered microwave weapon that would disintegrate all the electronics and make it impossible to isolate the fires or regain control of the system. As a result, emergency response teams would be overwhelmed, thousands of people would be killed or injured, gasoline rationing could be at World War II levels for months, and the entire economy would experience horrendous repercussions.

PRESIDENT CLINTON LENDS A HELPING HAND

In October 1995 the Defense Science Board Task Force on Information Warfare reported:

> Because of its ever-increasing dependence on information and information technology, the United States is one of the most vulnerable nations to information warfare attacks.[59]

Notwithstanding this stern warning from the Defense Department, in January 1996 President Clinton lifted the ban on exporting supercomputers to China. Within two years at least six hundred of the versatile machines had disappeared into the PRC. These are the very machines the unanimous, bipartisan Cox Committee stated could be valuable to the PLA's information warfare strategy and its associated munitions programs, including high-powered microwave weapons.[60]

In 1997 the President's Commission on Critical Infrastructure Protection reported that our civilian economy—energy,

banking, transportation, vital human services, and telecommu-
nications—is threatened by information warfare attack.

In other words, the civilian side echoed the warning Defense
had given two years earlier—the one Clinton ignored.

President Clinton's decision to allow the very central nervous system of information warfare to be exported to the PRC
was thus sandwiched between two of the most severe warnings
about America's vulnerability to offensive information warfare, warnings he did not heed.

For the first time since Appomattox, war, if it comes, will be
on the home front. Thanks to the Clinton administration, every
American is now at risk.

CHAPTER 9
THE SPRATLYS:
SECURING STRATEGIC GROUND

In the Spratly Islands... we fight for our national honor and military might.

—PRC newsletter

I t was blistering hot on the tarmac in Manila as the American congressional delegation boarded the plane for a reconnaissance mission over potentially hostile territory. The passengers aboard the Philippine C-130 Hercules that day in late 1998 were Republican Congressman Dana Rohrabacher, a respected legislator from the California coast; Al Santoli, his no-nonsense national security adviser; and Jeff "Skunk" Baxter, a Rohrabacher adviser best known as a guitarist with the Doobie Brothers rock band.

The mission: to expose the PRC's rapid military buildup in the strategic Spratly Islands in the South China Sea.

The Rohrabacher team had the aid of Philippine Congressman Roilo Golez, a U.S. Naval Academy graduate, and Alex Melcher, a trusted adviser to Philippine President Joseph Estrada.[1] The men considered the enterprise so important that they were willing to defy the wishes of the U.S. Department of

State, which wanted to avoid a confrontation with the Chinese Communists. Flying over the PRC's military facilities would be risky, but Rohrabacher felt that "any danger of retaliation by Chinese gunboats to our overflight was well worth the risk of informing the world of this aggression by the Chinese Communists that the Clinton administration is trying to ignore."[2]

But why was Clinton's State Department trying to halt a mission that could prove vital to the national security of the United States and our allies in the Pacific?

In State's opinion, the Rohrabacher mission would provoke the PRC. The State Department's cable read, "We do not believe it would be in the U.S. interest for the Congressional Delegation (CODEL) to travel to Palawan [the Philippine province closest to the Spratlys], since the presence of a highly visible senior USG [United States government] official on the island at this time would be taken as signifying more U.S. involvement in the current dispute between the Philippines and China over Mischief Reef than U.S. policy supports."[3]

Why was Clinton's State Department trying to halt a mission that could prove vital to the national security of the United States and our allies in the Pacific?

In Santoli's view, State's opposition was part of a larger pattern of appeasement toward the Chinese, one that allowed the PRC's unprecedented military buildup and power projection in the region. Although the State Department had tried to block Rohrabacher's efforts on other occasions, State's opposition to this mission was extraordinary.

This type of opposition could be coordinated only by the White House. It was no secret that the Rohrabacher team, along with others, had made life miserable for Communist China on Capitol Hill. Rohrabacher was a leader, along with Representatives Gerald Solomon (R-New York) and Nancy

Pelosi (D-California), in pushing to deny Most Favored Nation status to the PRC. He had helped uncover the illegal transfer of missile technology to China—made by corporations that were major Clinton donors. In case after case, Rohrabacher had opposed the Clinton administration's China policy.[4]

But still, for State to claim that it would not be in the interest of the United States for a member of Congress to witness a major PRC military buildup alongside a strategic naval passage was, in Santoli's opinion, "beyond mind-boggling and way over the line."

The Spratlys are a strategic island chain sitting astride vital shipping lanes to Japan, Taiwan, and South Korea, and offering access to developing oil and gas fields in the South China Sea.[5]

The Philippines have the strongest claim to the islands by virtue of proximity, but other nations make territorial claims as well, including Vietnam, Indonesia, and even the PRC, more than eight hundred miles to the north. (The Philippine government was indeed surprised in early 1995 when it discovered that the PRC had taken over territory in the Spratlys.)

As the C-130 pilot descended to five hundred feet, the crew saw three PRC naval vessels and six transport ships in the harbor. On the islands, legions of laborers and PLA engineers were frantically building a three-story reinforced concrete helicopter landing pad[6] and military fortification with gun ports—right in the heart of some of the world's most strategic real estate. It was obviously a military buildup of major proportions. Cameras clicked, capturing the military facilities on film. Although Rohrabacher and his men could clearly see PRC sailors scrambling to their anti-aircraft cannons—and although they were well aware of a direct threat from Beijing that "any reconnaissance flights by Philippine aircraft less than five thousand feet over Mischief Reef might trigger an accidental confrontation"[7]—the C-130 escaped unharmed.

As Congressman Rohrabacher later told the authors, "This is an historic and fateful move by China to bully the Philippines and other Asian countries who are allies of the U.S. The PRC effort to dominate the South China Sea and its critical passages to the western Pacific, Taiwan Straits, and Indian Ocean is an ominous threat to the United States and its interests in the Asia-Pacific region."

Unfortunately, this "ominous threat" has resulted from the Clinton administration's appeasement. The PRC was constructing a military facility almost a thousand miles from its mainland, and President Clinton's foreign policy team didn't want anyone to know about it. But Rohrabacher's team had gone ahead with the mission—and captured on film the heavily armed PRC navy gun ships and military fortification.

The move to fortify the Spratlys showed that Communist China is a dominant regional power and an emerging superpower.

Back in the United States, the photographs made front-page news. In spite of the usual PRC attempts at denial[8]—the Chinese officially stated that the construction was for refuge from storms for Chinese fishermen—here, finally, was direct proof that the PRC was moving from a defensive to an offensive position. Even a PRC newsletter profiling navy coastal troops pointed toward the real ambitions of the Chinese: "In the Spratly Islands... the Navy officers and men defend our southern border.... In their eyes, these reefs are a sacred part of their 'native land'.... We fight for our national honor and military might."[9]

The military buildup in the Spratly Islands signaled a fundamental shift in the military policy and capabilities of the PRC. To Rohrabacher, the move to fortify the Spratlys showed that the PRC was a dominant regional power and an emerging superpower. The PRC was on its way to possessing both the

Representative Dana Rohrabacher's team aboard the C-130, about to uncover the PRC's military buildup in the Spratlys. From left to right: Jeff "Skunk" Baxter, Rohrabacher, and Al Santoli.

military might and strategic positioning to dominate Asia in the next century.

With the president of the United States greeting—and taking money from—agents of the Er Bu (the PRC's military intelligence service) in the White House,[10] it was no surprise that Clinton's State Department had tried so adamantly to stop the mission and continue a policy of appeasement.

But despite the Clinton administration's best efforts to downplay the situation, the PRC has made its aggressive ambitions clear.

The buildup in the Spratlys is just part of the PRC's strategic power projection. A PLA-affiliated Chinese shipping company has grabbed control of major ports on both ends of the Panama Canal. As a result, the Chinese company could deny American ships access to the canal. The United States, which is the number one user of the Panama Canal, could see its trade crippled if this strategic shipping route is cut off. As Senate Majority Leader Trent Lott (R-Mississippi) wrote in a letter to Defense Secretary William Cohen, "It appears that we have given away the farm without a shot being fired."[11]

THE PRC: BUILDING FOR THE NEXT CENTURY

At the beginning of the twentieth century, as General Leonard Wood and Admiral George Dewey were arguing over how to position the American army and naval forces in the Pacific, American war planners saw Japan as the only potential enemy;[12] China was a spent force whose main fleet had been destroyed in the Sino-Japanese War of 1895.[13]

> The Clinton administration has enabled the People's Republic of China to smokescreen its growing naval capabilities.

At the dawn of the twenty-first century, a free and vibrant Japan is our ally, and the People's Republic of China is building up its naval forces and occupying strategic real estate astride sea lanes of commerce. As early as 1995 *Jane's Intelligence Review* reported that the PRC navy had a Spratly Task Force consisting of destroyers, frigates, patrol craft, submarines, and amphibious ships.[14]

The U.S. Navy could sweep such a force from the ocean, though it would have some trouble with PLA mines.[15] But with the advent of conventional air, surface, and subsurface cruise missiles, an engagement would be costly, and in a few years the American navy will be increasingly vulnerable.[16] A Russian Sovremenny-class destroyer with nuclear-armed cruise missiles makes a PRC naval task force even more deadly.[17] All naval planners know that a conventional cruise missile attack from 360 degrees—air, surface, and subsurface—is a deadly threat to a carrier battle group; the PRC leadership knows this, too.[18]

Since the early 1990s the PRC has been preparing a strategy of world military supremacy, according to the PLA's most senior military official, Liu Huaqing, who has an extensive naval background. In 1996 General Liu sent his daughter, Lieutenant Colonel Liu Chaoying, who also has a high-tech naval background, to offer money to President William Jefferson Clinton.[19] General Liu's words are eye-opening:

> Since the early 1990s, we have been adjusting and preparing for the world strategy of supremacy and power politics. Practice has proved that Deng Xiaoping's strategic judgement and Party Central Committee decisions were timely and farsighted, having laid certain grounds for China to resist supremacist military threats, blackmail, and aggression, and giving us time to upgrade our military equipment and develop and seize high-tech strategic ground.[20]

The PRC has devoted much of its high-tech military upgrade to its naval forces, since China has a mixed record as a sea-fighting nation.

The PRC's navy consists of five major arms—water-surface vessel units (destroyers, frigates, guided-missile boats, and various service vessels), submarine units, aviation units, coastal defense units, and marine corps. (See Appendix A for a detailed account of the PRC's current military balance.) From humble beginnings—a "speed boat" school in the early 1950s, its first four destroyers in 1954, its first nuclear-powered submarine in 1974, and its first missile launch from a submarine in 1982—the Chinese navy has grown considerably. Although, like most military forces, the PLA navy has experienced growing pains—including a mutiny[21]—in recent times the PRC naval modernization programs have begun to threaten the U.S. Navy.

Moreover, the Clinton administration has enabled the PRC to smokescreen its growing naval capabilities. In 1997, for example, Navy Secretary John Dalton denied to Congress that the PRC's China Ocean Shipping Company (COSCO) merchant vessels were even a *potential* tactical or strategic threat to the U.S. Navy. Dalton's denials notwithstanding, COSCO is a threat, as the Cox Report makes clear. The House Task Force on Terrorism and Unconventional Warfare revealed that "COSCO is actually an arm of the Chinese military establishment," and open sources tell us that in times of military crisis the PLA navy will mobilize COSCO ships. But the Clinton administration continues to deny that COSCO is a potential military threat; as the Cox Report points out, "The Clinton administration has determined that additional information concerning COSCO that appears in the Select Committee's classified Final Report cannot be made public." Any discussion of a PRC naval buildup is vulnerable to being officially sandbagged by PRC apologists such as Navy Secretary Dalton and National Security Director Sandy Berger.[22]

Nevertheless, others are looking closely at the PRC's military modernization. In December 1998 *Jane's Defense Weekly*, the open-source bible for military technology and trend spotting, put out an issue devoted to China and the road ahead. *Jane's* notes Communist China's plan to modernize by "the steady acquisition of locally produced warships and submarines augmented by the import of more sophisticated and powerful vessels and armaments from Russia and former Soviet states."[23] Communist China's overall goal is to limit freedom of the seas: "Navy chiefs are seeking to boost offensive sea denial capabilities with the purchase of heavily armed Sovremenny-class missile destroyers from Russia and indigenously developed Song-class and Russian KILO-class conventional submarines."[24]

Still, *Jane's* reports, "the PLAN's [People's Liberation Army Navy's] effectiveness has suffered because of poor training and a lack of experience in running modern warships."[25] *Jane's* is also skeptical of the PLA navy's ability to develop an effective submarine force but does conclude, "The Chinese leadership was acutely embarrassed by its inability to react to the deployment of two U.S. aircraft carrier battle groups in the Taiwan Strait and the PLAN was subsequently ordered to acquire the capabilities to prevent similar deployments by the U.S. Navy in the future."[26]

Unfortunately, it seems the PRC is rapidly addressing any perceived weaknesses. In May 1999 the Russian navy commander, Admiral Kuroyedov, said he was preparing to offer China direct technical support and "logistical support,"[27] a critical area where the Chinese are deficient. And, contrary to *Jane's* observation from the previous year, Admiral Kuroyedov praised the PLA navy's excellent training of ship commanders, which "is up to modern standards."[28] Perhaps the PRC's navy had made great strides in training since 1998.

The PLA navy has also made some progress on developing a credible submarine force. The Cox Report points out that the PRC is building "Boomers"—that is, intercontinental ballistic missile–capable nuclear submarines—and is working on nuclear warhead cruise missiles capable of being launched from a submarine's torpedo tube.[29] In addition, the PRC recently acquired Russian KILO-class diesel submarines.

Of utmost concern are recent reports that the PRC will purchase at least two intercontinental ballistic missile submarines (SSBNs) from the Russian government. It would take time for these SSBNs to become operational, but a PLA backed with SSBNs would pose a substantial threat to the United States and its military.

Still, the PLA submarine force has a long way to go—the U.S. Navy is a generation ahead in this field.[30] An official newsletter celebrating fifty years of the PLA navy lauded the "sacrifice" submarine crewmen make to patrol the waters; the newsletter described sailors who had "been out of port for *ten days* [emphasis added]" and who were "physically exhausted" and longing for their families. As any American submariner will tell you, ten days out of port does not constitute any sort of sacrifice; American submarine crews can spend most of the year at sea.

SPECIAL OPERATIONS—FROM THE SEA

An adjunct to the army until 1979, the PLA marine corps does not receive much attention. Nevertheless, an elite PLA "Frogman Unit" could be formidable. Another newsletter honoring fifty years of Communism in China depicts the Frogman Unit's daring special operations; it reads like an attack on an outpost in the Sprats or directly against Taiwan:

> Night has just fallen. The South China Sea is raging with roaring waves, surging to the skies. At the bottom of the sea, a submarine, resembling a giant whale, quietly approaches an "enemy-occupied zone." Soon afterward, from the waves emerge two camouflaged rubber dinghies, followed by eight fully armed men wearing diving apparatus, who then board a small boat. The small boat negotiates its way through the waves and makes a dash for the "enemy-occupied islet." Twenty minutes later the islet is shrouded in continual explosions and raging flames.

As with most elite units, the training to be a Frogman is arduous and dangerous. The unit's intensive training concentrates on all aspects of military training—weapons, communications, navigation, hand-to-hand combat, and so forth. It includes the most rigorous physical fitness standards as well as training in extreme conditions. As part of the selection process, Frogmen have to bury themselves in sand at temperatures greater than 100 degrees, lead a submachine gun attack in freezing cold mountains with only a cotton quilt and an overcoat, and survive alone for two weeks on a barren, uninhabited coral-reef islet with only a pack of biscuits, two packets of instant noodles, and a canteen of water.[31]

The unit's doctrine is simple: engage in feints to confuse the enemy, conduct harassment raids, carry out blockades, and strike and bomb the enemy's island (or coastal) facilities.

You can bet these warriors are active in the Spratly Islands and could be a key element in an invasion of Taiwan.

THE DUKE SPEAKS

Some sources believe that the PLA's air force and naval air branch cannot match America's air power. But the leading authority on the subject, Randy "Duke" Cunningham, believes it is extraordinarily dangerous to underestimate the PLA's air power.

Duke Cunningham, now in Congress representing California's 51st District, was without question the most accomplished U.S. Navy fighter pilot in the Vietnam War.[32] His rule for aerial combat is simple: "When I fight in the air I need to know my opponent's aircraft and weapon system capabilities, and I enter the fight believing he is the best fighter pilot in the world. In one turn I can tell if he is the best; if he isn't, I kill him quick."[33]

When Cunningham was in the navy after Vietnam, he was responsible for assessing real-world threats facing the U.S. Seventh Fleet, including the PRC and North Korea. To this day Cunningham flies high-performance navy fighters, and he has strong opinions on the emerging PRC threat, especially Beijing's military aircraft modernization program. According to Cunningham, "In a life and death struggle with the PRC they are willing to expend millions of lives and assets." Thanks to PRC spying, the Communist military leaders can spend more on upgrading their fighter force since they do not have to invest in developing strategic weapons systems. Cunningham states, "The PLA have capable weapon assets and are now acquiring further advanced technology that makes their air force a danger not only to the U.S. but to the rest of the world."[34]

That the PRC has modern aircraft and weapon systems is dangerous enough, but most disturbing is that, as Congressman Cunningham revealed, the Clinton administration has allowed PRC pilots to observe U.S. fighter tactics firsthand at Top Gun.[35] When planes and weapons are equally capable, training and tactics in aerial warfare are everything. In one brief moment in his Capitol Hill office, Cunningham, who is respected as a man of quiet authority, let his true fighter pilot nature show; he was white hot at the Clinton White House's irresponsibility in allowing PRC aviators to visit Top Gun. Many American naval aviators have paid with their lives to perfect successful fighter tactics. For the Clinton administration to allow PRC pilots to observe our maneuvers is nothing short of betrayal. Unfortunately, that betrayal is part of a larger pattern of misguided appeasement that compromises American security to raise campaign cash from foreign nationals and from domestic corporate interests that put profits ahead of country.

CHAPTER 10
SEIZING TAIWAN

Taiwan is the turd in the punchbowl of U.S.-China relations.[1]

—Admiral Dennis Blair, commander-in-chief, Pacific Forces

Taiwan—officially, the Republic of China[2]—lies about a hundred miles off the Chinese coast, roughly the same distance the D-Day forces traversed to storm the beaches at Normandy. Despite the proximity, a Chinese amphibious assault on Taiwan today would be even more difficult than the allies' invasion of June 6, 1944. The Taiwan Straits are known for bad weather; the PLA does not, as yet, control the seas and skies, nor does it have anywhere near the lift capacity the allies had; and Taiwan has few beaches on which the Chinese could land.

Playing golf at Linkou, high up on the cliffs overlooking the Taiwan Straits, one of the authors saw firsthand just how difficult a PLA amphibious assault would be. Like something out of *The Guns of Navarone*, the sheer cliffs are hundreds of feet high, heading directly down into a crashing sea. There is no beach.

Nevertheless, Taiwan cannot rest easy knowing that a PLA amphibious invasion would be difficult. The island nation faces other perils.

We believe that the most likely place for substantial PLA military action in the near future is in Taiwan.

In fact, the PLA may be prepared to use force against Taiwan quite soon. Asian newspapers have reported that in August 1999 PLA air and naval forces were put on combat-readiness status to increase pressure on Taiwan. A Beijing source revealed that the PLA was considering invading and temporarily occupying an outlying Taiwanese island in a show of force. The source stated, "Hardliners at Beidaihe urged that action be taken soon after the October 1 national holiday," referring to the fiftieth anniversary of the PRC.[3]

THE POWDER KEG

When the fortunes of war turned against the Chinese Nationalists on the mainland in 1949, they retreated to Taiwan. Mao Zedong expected to take Taiwan quickly, but the Chinese National Army held off the PLA, killing or capturing ten thousand of Mao's soldiers.[4] Then, when the North Koreans invaded the South in June 1950, President Harry Truman put the United States Seventh Fleet between Taiwan and the mainland coast, which generally stabilized the situation between the Communists and the Nationalists.

Taiwan is the most likely target for substantial military action by the People's Liberation Army.

Things began to change in the late 1980s. The PLA's massacre of the students at Tiananmen reminded the world of the illegitimacy of the Beijing regime. Taiwan's honest and able president at the time, Chiang Chingkuo, lifted martial law on the island, released dissidents, promised free and fair elections, and set the country on an irreversible course toward democracy. As the Communists on the mainland threw more and more Chinese patriots into labor camps, the contrast between Beijing and Taipei became stark.

By the spring of 1994 President Chiang had been succeeded by American-educated President Lee Teng-hui, the first native Taiwanese to become president of the Republic of China. In 1996 President Lee would become the first freely elected chief of a Chinese state.

Events came to a head in early May 1994 and went downhill from there. On May 2 President Clinton had a forty-minute meeting in the Oval Office with a particularly odious member of the Chinese Communist Party's Politburo who had married into a corrupt military family and had spent his career as an official with the arms smugglers.[5] Meanwhile, America's representative in Taiwan[6] was telling Taiwan's foreign minister that President Lee, who was on his way to South Africa for Nelson Mandela's presidential inauguration and needed to refuel his plane in Honolulu, could not set foot in the United States. Though the State Department frantically reversed course and allowed a refueling stop, President Lee, still uncertain of what the United States had ultimately decided, was offended by the treatment he had received and did not get off the plane in Hawaii.

When word reached Capitol Hill that Clinton had welcomed a Politburo member to the Oval Office while President Lee had been snubbed, Congress hit the ceiling—senators and congressmen, Republicans and Democrats alike. Later the Senate voted 94–0 demanding that the State Department grant visas to Taiwanese officials. Congress was further outraged when, shortly thereafter, President Clinton reversed his publicly stated position and "delinked" human rights issues from trade with the PRC. Democrats felt betrayed, and Republicans felt that Clinton was a hypocrite for having attacked George Bush's "coddling" of the PRC during the 1992 campaign. If Congress had known then that associates of Chinese intelligence had frequent access to the White House and the First Family,[7] the response would have been even stronger.

When in the spring of 1995 President Lee, a Ph.D. from Cornell, was invited to his university's reunion, the PRC voiced its objection.[8] For Congress the issue was simple: the president of a democratic country with a long history of friendly relations with the United States versus the Communist regime that killed the kids at Tiananmen. The House unanimously passed a resolution inviting Lee, and the Senate followed with a 97–1 vote.[9] Recognizing the overwhelming support for Taiwan, the White House allowed him in.[10]

Stepping up its pressure on Taiwan, the People's Liberation Army fired ballistic missiles off the island's coast.

This time, Beijing hit the ceiling. On July 21, 22, and 23, the PLA's Second Artillery fired a total of six M-9 ballistic missiles in the ocean about one hundred nautical miles north of Taipei. It was the first time in history that nuclear-capable missiles had been used in a hostile environment.[11] The PLA's Second Artillery fired and reloaded in full view of American spy satellites and then followed up with sea and air exercises that involved the test firing of cruise missiles.[12]

The effect on Taiwan and the region was immediate. More than a thousand airline flights had to be diverted or delayed. According to one estimate, $14 billion left the county within nine months.[13] Many of Taiwan's political and economic elite, who had American passports or residency permits abroad, fled the danger zone. One Taiwan politician complained privately that many of her friends had "grabbed their U.S. passports out of the safe and headed to the airport to visit the kids in L.A. [Los Angeles]."[14]

Surprisingly, there was little response from Congress. Only Senator Larry Pressler (R-South Dakota) took to the floor to denounce Beijing's nuclear intimidation of Taiwan.[15]

Perhaps because of Congress's laid-back response, Beijing was encouraged to raise the ante. Fairly quickly it became clear that (1) the PLA was going to try some new intimidation in 1996 and (2) the issue was whether the United States would intervene.

The PLA's chief spymaster, General Xiong Guangkai— already infamous for his role in the Tiananmen Square massacre—at one point threatened the United States with attack. Xiong sent his thinly veiled threat through former Clinton official Chas Freeman, stating that American officials "care more about Los Angeles than they do about Taiwan."[16] We now know that a Chinese intelligence agent walked into the CIA that same year with documents showing that Chinese spies had stolen nuclear secrets from America's Los Alamos National Laboratories.[17] The CIA later concluded that PRC intelligence had purposely directed the "walk-in" to the Americans, perhaps to demonstrate the PLA's capabilities.[18] According to news accounts, the Los Alamos operation was run by Chinese military intelligence, which is headed by General Xiong.[19] In short, Xiong knew he was holding the high cards when he spoke to Freeman. The message was clear: The PLA's Second Artillery could back up General Xiong's threat with "Made in USA" nuclear weapons targeted at American cities.

(Apparently the fact that the Clinton-Gore administration had welcomed General Xiong to Washington hadn't softened his hard-line stance toward the United States.)

In March 1996 the Second Artillery fired four more missiles, this time bracketing Taiwan's largest northern and southern ports about thirty nautical miles out to sea. The Chinese followed up with another massive sea exercise and with a mock invasion of an island with topography similar to Taiwan's. Television shots of thousands of troops storming ashore were shown over and over on Hong Kong television and on local television within range of Taiwan's offshore islands.

This time the United States responded. The Clinton administration sent the aircraft carriers *Independence* and *Nimitz* to the Taiwan Straits. But without the forceful bipartisan pressure applied by Representative Chris Cox (R-California), Representative Nancy Pelosi (D-California), and their colleagues, the carriers likely would not have been sent. Until it became clear that Cox and Pelosi were going to embarrass the Clinton administration in an election year, nothing was going to happen.[20]

After America's intervention, Taiwan had its free election, President Lee won, and the Chinese Communists backed off. There the matter rested, at least until the summer of 1999, when the PLA began to exert more pressure on Taiwan.

We believe aggression against Taiwan will return. But what form will this aggression take, and when will it occur?

WAR IN THE STRAITS

Three possible routes of PRC aggression against Taiwan are:

- A nuclear attack
- A D-Day type invasion
- A rolling blockade of the island

Beijing, a nuclear power since 1964, could use atomic weapons against Taiwan or threaten to do so using a single demonstration device. But that would be a catastrophe for both China and the Chinese Communist Party: Diplomatic ties with the rest of the world would be immediately cut off; the PRC would be subject to an international embargo; members of the regime and their families would be subject to arrest as war criminals if they left the country; any foreign property held by the regime or its supporters would be confiscated; every one of China's neighbors would begin a massive arms buildup. These negatives make a nuclear attack unlikely.

Chinese military planners have discussed the D-Day invasion scenario. In a 1993 study entitled "Can the Chinese Army Win the Next War?" the authors (presumably PLA navy officers writing under pen names) concluded that an armed invasion was not the best means of handling "the problem of the return of Taiwan."

The American military agrees. A spring 1999 Defense Department report to Congress, "The Security Situation in the Taiwan Strait," says, "An amphibious invasion of Taiwan by China would be a highly risky and most unlikely option for the PLA, only as a last resort to force total surrender of the island." The Department of Defense estimates that today the PLA navy has the sea-lift capacity to "transport approximately one infantry division," which would not be enough to take on Taiwan's ten infantry divisions, two mechanized infantry divisions, six armored brigades, and two marine divisions. Moreover, the PLA would have to account for Taiwan's navy, which has some new ships, and its air force, which is integrating 150 F-16s and 60 French Mirage 2000-5s into its inventory of 100 second-generation fighters.

Although an all-out PLA invasion of Taiwan accompanied by a sea-air battle over the straits would devastate Taiwan, the island nation's military would use its limited attack capability to

cause as much damage to the mainland as possible. Even after
a successful amphibious assault, Beijing would find a ravaged
island, wholesale destruction in south and east China, hundreds
of thousands of dead soldiers, and millions of wounded soldiers
and civilians on both sides of the straits.

The Department of Defense estimates that by 2005 the PLA
will be able to overcome the sea and air problems enough to
succeed in invading Taiwan—"barring third-party intervention"—but Beijing would have to be "willing to accept the
almost certain political, economic, diplomatic and military costs
such a course of action would produce." In other words, an
amphibious invasion would still be unlikely.

A third option is the blockade.[21] In this case, Beijing would
attempt to starve the island of Taiwan in the same way Hitler
tried to starve England with his U-boat campaign. The PLA
navy has a distinct edge over Taiwan in submarines, including
KILO-class subs purchased from Russia.

In Congress, Democrats and Republicans agree that a strong
Taiwan deters war in the Far East while a militarily weak Taiwan invites trouble from the PLA. But the Clinton administration's policy is the opposite: A strong Taiwan is provocative.
Consequently, the administration has so far refused to sell Taiwan the American-built submarines it needs to defend itself
against a blockade, even though the proposal was first made in
Congress as far back as 1995.

A blockade, accompanied by violations of Taiwan's air space
and mining of harbors, could succeed, but only *over time*. Non-
Taiwanese vessels would not risk becoming entangled due to the
resulting insurance rates, and eventually Taiwan would have
no choice but to surrender.

The risks are still too steep for Beijing. Since the blockade is
unlikely to succeed quickly, it would give Taiwan supporters

time to bring the United States into confrontation with the PRC. The U.S. Navy would not have to run the blockade; the United States and the other democratic powers would simply have to impose a counterembargo on the PRC. Beijing can't make up in trade with North Korea and Cuba what it would lose in Europe and the Americas.

Whatever plan the Communists try in order to recover Taiwan, it *must* succeed. If they try and fail, millions of Chinese will question whether the regime can be beaten. After Tiananmen, the Communists can't risk that.

In order to give the go signal, Beijing would have to be convinced that the plan would have:

- A high likelihood of success
- A low likelihood of collateral damage
- A low likelihood of outside intervention

This can be done.

AN ELECTRONIC PEARL HARBOR

It is our military judgment that the greatest threat to Taiwan is what might be called an electronic Pearl Harbor.[22] The elements of this threat are as follows:

- Surprise attack
- Information warfare
- Psychological warfare
- Precision-guided munitions
- Denial and deception
- Subversion
- Special operations

The PLA is hard at work on all of these elements; it even claims to have invented some of them. The Chinese are beginning to incorporate them into their training exercises. By contrast, Taiwan is woefully behind on all of these elements and, in some cases, not aware of them at all.

SURPRISE ATTACK

In its latest report to Congress on PLA military capabilities, the Department of Defense noted that PLA doctrine "includes the option of preemptive military action"[23]—Pentagonese for surprise attack. The PLA's major daily expressly endorsed this doctrine in 1996,[24] and PLA military planners have, for instance, noted the advantage of defeating an enemy with a long-range, surprise air attack.[25] The principle would be to catch Taiwan's navy in port and its air force on the ground, and to strike so quickly that the United States and the other democratic powers would not have time to react. Sophisticated early warning radar would enable Taiwan to fend off a surprise attack, but the Clinton administration has stalled and not let the Taiwanese have this equipment.

INFORMATION WARFARE

As we have seen, the PLA has been actively developing its information warfare capabilities. Information warfare has already been used effectively; in the Gulf War, the allies slipped a computer virus into the command and control computers for Iraq's air defenses. Today, offensive information warfare is far beyond the point of simply shutting things down. *Manipulation* of the enemy's systems can cause much more damage. For example, it may be possible to send a surge through the opponent's electric power grid that will melt anything attached to it. And why shut down banks when you can raid the accounts of the enemy's political and economic elite, sending the money overseas or wiping out any record of the accounts? Another option: Allow the rich to transfer their money overseas and then publicize it. The water and sewer systems would be other obvious targets, as would modern, computerized factories and refineries, air traffic control systems, communications, and even traffic lights.

What is clear is that the PLA has completely embraced the idea of information warfare—Chinese military writings speak of "Tak[ing] Information Warfare as a Starting Point of Military Struggle Preparations."[26]

PSYCHOLOGICAL WARFARE

> The mind of the enemy and the will of his leaders is a target of far more importance than the bodies of his troops.[27]
>
> —Mao Zedong

In recent years the PLA has produced articles, in publications it knew Taiwan would read, ridiculing Taiwan's defense preparations. More psychologically effective were the missile tests the PLA conducted in 1995 and 1996.

But the Chinese army would gain the greatest psychological edge were it to use information warfare against troops or the home front. The first target of offensive information warfare would be the opponent's command and control network, which is good only if the troops have faith in it. The first time an opponent successfully breaks into the system and uses it to issue false orders, paralysis will develop; even with redundant backups, if one order is suspect, they're all suspect. Therefore, PLA hackers would not have to break into every system; word getting around of one PLA success—or even the *false* report of a PLA success—might be enough. Troops would be reluctant to trust their own sensors, and they would be equally reluctant to carry out orders that might lead them into an ambush.

On the home front, the ability of information warfare to manipulate and destroy systems can be deadly for civilian morale—particularly when the nation's political and economic elite begin fleeing the country.

PRECISION-GUIDED MUNITIONS

Taiwan's most significant vulnerability is its limited capacity to defend against the growing arsenal of Chinese ballistic missiles.[28]

—United States Department of Defense, 1999

As it demonstrated in 1995 and 1996, the PLA has a significant capability in modern, mobile, ballistic missiles. The Department of Defense estimates that the PLA will have six hundred missiles opposite Taiwan by 2005.[29] These missiles are becoming more and more accurate thanks to improvements in on-board systems and the use of the Global Positioning System (GPS). According to the Cox Committee's report, "The PRC has stolen U.S. missile guidance technology that has direct application to the PLA's ballistic missiles and rockets."[30] The Cox Committee also reports that space/rocket technology obtained from America's own Hughes Corporation makes PLA missiles more reliable.[31]

The Chinese have built a full-scale mock-up of Taiwan's largest air base and are reportedly practicing assaults on the facility.

Cruise missiles are another PLA strength and corresponding Taiwanese vulnerability. The PLA began with Soviet types, moved on to reverse-engineered French models, and is working on its own supersonic versions. The PLA has a major effort going on to produce a long-range land attack cruise missile similar to the American Tomahawk,[32] and Russia is providing stealth technology to the Chinese cruise missile program.[33]

A surprise attack against Taiwan would begin with ballistic and cruise missiles in a barrage designed to overwhelm command and control centers, blind radars and other warning systems, sink Taiwan's navy in port, and put military airfields out of commission.

DENIAL AND DECEPTION[34]

All warfare is based on deception.[35]

—Sun Tzu

Denial would include attacks on Taiwan's sensors and warning systems and possibly an attempt to blind American spy satellites. Beijing is known to be working feverishly on land-based anti-satellite weapons systems.[36] These systems may be based on high-powered microwave weapons technology Chinese spies stole from the United States.[37] According to one unconfirmed report in Washington, such a weapon was recently tested successfully against a European satellite system.

Deception, as part of an information warfare attack, would include false reports of enemy troop movements and tainted orders designed to bring the command and control system to a halt. The key to deception, as we have seen, is to create chaos by breaking the troops' faith in the command and control system.

SUBVERSION

Four former citizens of Taiwan—John Huang, Johnny Chung, Maria Hsia, and Charlie Trie—have been indicted or convicted for their roles in funneling illegal campaign contributions to the Democrats. Huang took the Fifth Amendment more than a thousand times when asked by Judicial Watch if he had ties to Chinese intelligence; Chung has confessed his own role to the federal grand jury; Hsia worked as a knowing PRC agent, according to the CIA;[38] and Trie was the bagman for a Macau criminal syndicate figure with his own ties to the PLA. Moreover, the three scientists[39] accused in the press of transferring American nuclear secrets to the PLA were former citizens of Taiwan. The full measure of the damage their activities caused to Taiwan's security won't be known for years, but it is substantial.

Taiwan's internal security forces (and the FBI) could have a serious problem, because it is unlikely that the Magnificent Seven were the only ones subject to offers of money or appeals to vanity.[40] The Communists could certainly co-opt some political figure on Taiwan, put him or her on a piece of the island they've seized, and then let their puppet rescind any calls for help from the United States. Credible news reports reveal that the PRC already has more than 34,000 Taiwanese intelligence operatives.[41]

Another headache for Taiwan would be continued PRC subversion in the United States. It has recently been revealed that many years ago the Soviet KGB owned a congressman from New York.[42] PRC intelligence, having been so spectacularly successful in the Executive branch, can be expected to target Capitol Hill, Beijing's most effective opponent. According to one unconfirmed report, the wife of Beijing's deputy chief of mission in Washington is in charge of this operation, which would target senators and members of Congress as well as key staff members, particularly those with national security responsibility. Money, in the form of after-government job prospects with pro-Beijing business interests, and appeals to personal vanity would be the most likely inducements.

SPECIAL OPERATIONS

Satellite photos (shown in this book's photo section) reveal that the Chinese have built a full-scale mock-up of Taiwan's Ching-chuankang (CCK) military airfield. CCK, Taiwan's largest air base, was designed by the American Strategic Air Command to handle B-52 bombers and was an active U.S. Air Force base during the Vietnam War. Today it houses several fighter wings of Taiwan's air force. Apparently PLA paratroopers and special forces have been practicing assaults on the mock-up, which includes exact duplicates of CCK's runways, fuel storage depots, and aircraft revetments.[43] The mock-up was undoubtedly in place for some time without being discovered. We can only wonder how many other Taiwan military facilities have been duplicated and not yet been found.

Creating an exact replica of something as big as CCK is very expensive. The PRC's Finance Ministry must be convinced the project is a priority—a pretty ominous thought. In the event of a successful PLA attack on CCK, it could serve as a headquarters for the PRC's temporary puppet regime.

THE TAIWAN RELATIONS ACT

The United States Congress was outraged when President Jimmy Carter recognized Beijing, severed diplomatic relations with the Republic of China, and canceled our defense treaty with Taiwan. In response, a Congress controlled by the Democrats imposed the Taiwan Relations Act on Carter by a veto-proof margin. The act dictates that:

> The President is directed to inform the Congress promptly of any threat to the security or the social or economic system of the People of Taiwan and any danger to the interests of the United States arising therefrom. The President and the Congress

shall determine, in accordance with constitutional processes, appropriate action by the United States in response to any such danger.[44]

In other words: If Taiwan is attacked, we're coming.

Beijing, consequently, is concerned with how to keep the Americans at home. The answer could be the Sovremenny.

THE SOVREMENNY

Watch your feet as you step aboard a Russian Sovremenny-class destroyer. If you don't, you may take a nasty fall over the tracks in the deck. The Sovremenny is armed with forty sea mines that can be rolled on these tracks and off the stern. American and British destroyers don't carry sea mines.

Standing on the fan tail of a 7,300-ton Sovremenny, the observer is struck by its sheer offensive power: eight nuclear-tipped SS-N-22 Sunburn anti-ship missiles, four five-inch guns (NATO destroyers typically have one), nuclear-armed wake-homing torpedoes, anti-submarine mortars, and batteries of anti-aircraft missiles and guns. There's no effort to make this a stealthy ship; it's almost as if the PLA doesn't care if it's seen on radar. Lines and gear that seem to have no purpose stick out all over the ship. To the visitor on board the destroyer, it's not clear if the hatches are water-tight or if the crew has had any serious training in damage control. The ship would be in real trouble from incoming American Harpoon anti-shipping missiles. For the crew of a Sovremenny, it's strike first or die.[45]

The Sovremenny and its SS-N-22 missile system were designed to do one thing: kill American aircraft carriers and Aegis-class cruisers. The SS-N-22 missile skims the surface of the water at two-and-a-half times the speed of sound, until just before impact, when it lifts up and then heads straight down

This photo, taken by one of the authors, shows a Russian Sovremenny-class destroyer. The heavily armed Sovremenny will help the PRC counter America's naval forces.

into the target's deck. Its two hundred–kiloton nuclear warhead has almost twenty times the explosive power of the atomic bomb dropped on Hiroshima.[46] The PLA navy has at least two Sovremennys on order from Russia.[47]

The U.S. Navy has no defense against this missile system. Eight nuclear-tipped[48] SS-N-22s fired in barrage at an aircraft carrier battle group would result in the immediate death of thousands of American sailors, airmen, and on-board marines. It's unlikely that anyone would survive.[49]

In the hands of the PLA navy, the Sovremenny is designed to be a high-stakes deterrent. The message to the U.S. Navy is clear: Stay away. As retired Rear Admiral Eric McVadon put it, "It's enough to make the U.S. 7th [Pacific] Fleet think twice."[50]

TAIWAN'S RESPONSE

In 1995, a decade after the PLA started on information warfare, Taiwan's military establishment was complacent: "As far as our side is concerned, the defense of the Taiwan Strait area is mainly a matter of resisting landing operations."[51] As recently as the fall of 1998, the Taiwanese had given virtually no thought to information warfare.[52]

But there has been some change of late. At a March 1999 conference, Taiwan's minister of defense, Tang Fei, surprised the audience by indicating Taiwan may be powerless against information warfare. A small amount of money ($4 million) has now been allocated to begin working on network security programs. Though it's not much, at the very least the man at the top realizes there's a problem.[53]

Likewise, Dr. Lin Chong-pin, vice chairman of Taiwan's Mainland Affairs Council, has been spreading the alarm about Beijing's new capabilities. "Our psychological defense is almost nil," he told reporters in the spring of 1999.[54] Dr. Lin has a particularly fine reputation as an analyst since he correctly predicted that Beijing would use its new missiles to try to intimidate Taiwan in 1995.[55]

THEATER MISSILE DEFENSE

Theater Missile Defense is currently the most controversial security issue among the United States, the PRC, and Taiwan. The United States is discussing the system with Japan and South Korea, both of which are concerned about North Korea's missile program. Taiwan, however, is in a quandary. The United States Congress favors including Taiwan in a Theater Missile Defense system, but Beijing is adamantly opposed. And the Clinton administration hopes the issue will go away.

Taiwan recognizes that missile defense is, in the words of the U.S. Defense Department, its "most significant vulnerability."

The government is already developing the American Patriot 3 air defense system, which should give the island some limited protection. Theater Missile Defense has been tested successfully, but deployment is still years away. Even if successful, a Theater Missile Defense program would be very expensive, so much so that Taiwan might have to crowd out other important defense programs or even civilian spending.

Beijing and its friends have been screaming that building a Theater Missile Defense system for Taiwan would be "provocative." Actually, the reverse is true. Beijing's missile buildup and bullying tactics are provocative, while Theater Missile Defense is a purely defensive system. It would deter a Communist dictatorship's designs on a democratic country—which is why Beijing is unhappy.

Theater Missile Defense also reduces the likelihood that Taiwan will build its own missile-based nuclear counterstrike force. Taiwan has so far refrained from building up its missile systems, but, with the rising PLA missile threat, public discussion of the issue has grown.[56] Elected officials on Taiwan have told the authors that they are beginning to get "Let's blast 'em!" comments from constituents on radio talk shows. If this sentiment grows, it could cast pressure on Taiwan's democratically elected officials.

"HAPPY NEW YEAR!"
A nightmare for Taiwan might look like this:[57]

■ The surprise attack would come at Tet, the Chinese New Year, which usually falls in February on the Western calendar. This is what the North Vietnamese and Viet Cong did in 1968. The PLA would know that many of Taiwan's military people would be on leave and that readiness would be at its lowest point of the year. Moreover, if the attack fell during the U.S. Congress's February recess, the United States would have difficulty responding quickly.

■ The Chinese, having stationed their Sovremenny destroyers around Taiwan, would inform the U.S. State Department to stay out of the conflict.

■ The PLA's precision-guided ballistic and cruise missiles would simultaneously hit Taiwan's navy and air force as well as selected ground targets.

■ The PLA would launch a massive information warfare assault on the island. Some command and control networks would be destroyed while others would be deliberately spared so they could be manipulated from the inside.

■ All radio and television signals would be jammed, and PLA hackers would broadcast a false image of President Lee ordering the military to surrender.

■ The PLA would not shut down the banking system, but instead it would so scramble the accounts that they could never be reconstructed. Meanwhile, the wealth of Taiwan's political and economic elite would be transferred to Communist-controlled accounts abroad.

■ The power grid would be left in place until the information warfare campaign had done its work, and then a surge of electricity would blow out the electrical grid.

■ Taiwan's industrial base—factories, refineries, nuclear power stations, communications facilities—would be corrupted or destroyed.

■ One civilian airport would be deliberately left open so that the hope of escape would contribute to the panic and chaos among the population.

■ The Chinese information warfare campaign would deliberately leave some Taiwanese radar systems intact to warn of "virtual assaults," feeding the confusion and bringing the command and control system to a halt.

- Fifth columns at home and abroad would spread rumors and try to keep Washington confused and divided.
- Special forces would take over an air base. A well-known Taiwanese political figure who had been co-opted by the Chinese would send a message to Washington rescinding any of Taipei's calls for help.

It would all be over in seventy-two hours.

CHAPTER 11
THE THREAT TO JAPAN

The PRC and DPRK [North Korea] are... in the relations of
lips and teeth, and the peoples and armies of the two countries
have a blood-tied traditional relationship.[1]

—General Xiong Guangkai, 1999

On the last day of August 1998 Japan awoke to find
that its security assumptions had suddenly and
radically changed, and not for the better: North
Korea had sent a three-stage intermediate-range ballistic missile roaring across Japan's northernmost island. Just as the Iranian missile test the previous month had been unexpected, the
American officials on whom Japan has depended for fifty years
were caught off-guard. An American people preoccupied with
the Monica Lewinsky scandal didn't notice, but this missile test
caused a sensation in Japan—and its effects still linger.

The immediate question was, how could a poverty-stricken
Third World country like North Korea[2] produce such a sophisticated—and expensive—device?

The answer, at least in part, is the "Made in China" label
figuratively stamped on the side of the missile.

JAPAN'S NIGHTMARE: A NUCLEAR-ARMED NORTH KOREA

In 1997 the PRC Foreign Ministry inadvertently revealed that the PRC and North Korea had signed a secret agreement calling for Beijing to supply "military assistance" to Pyongyang.[3] Based on General Xiong Guangkai's frequent contact with the North Korean military, we believe that Xiong, the PRC's chief military spymaster, is in charge of this program.[4] We also believe the PLA is directly assisting the North Korean missile program and using North Korea as a cover to ship PLA missiles to terrorist destinations in the Middle East.

During the late winter of 1994, an American Defense Intelligence Agency analyst discovered startling evidence of direct PLA assistance to North Korea's missile development. When the North Koreans rolled out a test model of their latest ballistic missile, the analyst, working off satellite pictures, put an overlay of an older Chinese missile, the CSS-2, on top of a photo of the new North Korean missile—and the rivets matched. The first stage of the North Korean test model and the CSS-2 were virtually identical. The Defense Intelligence Agency official concluded, "Presumably, the only way they [North Korean engineers] would know how to build something the size of the CSS-2 is either by physical transfer of such a beast, or of engineers familiar with the program."[5]

This 1994 test model became the operational Taepodong I missile that flew over Japan four years later. The Chinese CSS-2 was designed as a nuclear weapon; about forty of these missiles are targeted on neighboring Asian countries.[6] Clearly, the Taepodong I is designed to carry a nuclear or possibly biological[7] warhead, which the Chinese government would have known when it transferred the technology to North Korea.

In the summer of 1999, Secretary of State Albright was running for cover after yet another China proliferation report in the

Washington Times. "We are concerned by reports that the DPRK [North Korea] may be seeking from China materials such as specialty steel for its missile program," she said.[8] The previous day, the *Times*'s Bill Gertz, quoting American intelligence sources, revealed that PLA companies were secretly transferring specialty steel, accelerometers, and gyroscopes to the North Korean missile program. The manufacturer of these parts would have been Chinese spy Lieutenant Colonel Liu Chaoying's former employers, the China National Precision Machinery Import-Export Corporation. When pressed, Albright said the Clinton administration "will fully and faithfully implement the requirements of U.S. law."[9]

But the administration never took any action against the Chinese arms dealers.

Actually, the PLA–North Korean missile connection is not much of a secret to specialists who watch this sort of thing. According to a 1995 South Korean account, China's strategic weapons agency, COSTIND, is training between fifty and two hundred North Korean missile engineers. Seoul also reports that North Korean No-dong missiles have primary components almost identical to those of the CSS-2.[10] According to American proliferation expert Gary Milhollin, in the 1970s and early 1980s Chinese firms "provided technology for rocket engine design and production, metallurgy and airframes" to North Korea. They also helped reverse-engineer and upgrade Soviet-supplied Scud missiles, which North Korea later sold to Iran and Syria.[11] Another American specialist, Dr. Seth Carus, believes North Korean cruise missiles "probably depend on imported Chinese components, especially the sustainer motors and guidance."[12]

When he announced his report on Chinese espionage, Representative Chris Cox (R-California) took special note of the

PRC as a "significant proliferator" and predicted that national security information stolen from the United States may end up "in the hands of regimes much less stable than the People's Republic of China."[13] The National Security Agency has already intercepted communications revealing that the PLA is sharing space technology with North Korea.[14] And we know from the Loral-Hughes affair—when the two American corporations gave the PLA sensitive satellite technology—that space launch technology has an immediate application to ballistic missile programs. It is thus conceivable that the space launch technology passed to North Korea might, at least in part, have originated in the United States.

It seems, then, that North Korean arms programs, particularly missiles, are adjuncts of China's programs. As the Chinese replace older missiles, they could transfer to North Korea the technology and equipment to make the older missiles.

Chinese arms dealers could even be using North Korea as a clandestine transshipment point for Chinese-origin arms sales to the Middle East. Quoting U.S. intelligence sources, the *Washington Post* in 1987 reported that the Chinese were doing just that with Silkworm missile deliveries to Iran.[15] Similarly, in February 1989 *New York Times* columnist William Safire revealed that the Chinese had restricted the travel of U.S. diplomats in Shenyang, Manchuria, not far from the Yalu River, which separates China from North Korea; Safire speculated that the Chinese did not want American diplomats to see Chinese missiles crossing the Yalu and being loaded on North Korean ships bound for Iran.[16]

China also serves as an illicit conduit for Western arms production equipment to North Korea. In January 1994 the Japanese police raided the Yokohama Machinery Corporation and the Aritsu Corporation, an affiliate of electronics giant

NEC. Both were suspected of having illegally exported spectrum analyzers, which can be used to improve missile guidance systems, to North Korea via China.[17]

PLA weapons cooperation with North Korea benefits both countries economically and politically. In North Korea, the PLA has found a market for its older missiles, meaning some uncompetitive Chinese factories can remain open to produce parts for the older generation of missiles. Because of its concern over trade relations with the United States, Beijing is always looking for deniability, and North Korea provides the Chinese an indirect means of exporting missiles to countries such as Libya. Exporting Chinese missiles, in turn, allows North Korea to repay the Chinese for developing its weapons program under the May 1996 agreement.[18] In addition, by transporting Chinese-origin missiles to various hot spots, the North Korean merchant marine gets a bit of business. In the summer of 1999 Indian authorities detained a North Korean ship bound for Pakistan and, according to the *Times of India*, found "machinery and parts to manufacture missiles of Chinese origin."[19] By one estimate, North Korea receives about $100 million per year in missile export earnings.[20]

The People's Liberation Army might have transferred to North Korea some or all of the nuclear warhead designs the Chinese stole from the United States.

How involved are the Chinese in North Korea? In 1985 one of the authors had an extended interview with the Chinese Foreign Ministry official responsible for North Asian affairs, who complained that the North Korean government was very successful in playing off the PRC against the Soviet Union. "The [North] Koreans are very difficult," he said.[21] His comment was probably accurate—at that time. But the Soviet Union no longer exists, and North Korea's longtime dictator, Kim Ilsung, has been dead for several years. North Korea's Communist regime now depends completely on China for food, fuel, and political support.

Today, the PRC's official position, as expressed to one of the authors,[22] remains what it was in 1985—"Sorry, can't help you in North Korea." But some members of the Chinese Communist aristocracy are probably making money in North Korean arms deals, just as they do with arms sales to Iran, Pakistan, and other trouble spots. In fact, General Xiong was in North Korea in August 1998, only a couple of weeks before the Taepodong I test across Japan.[23] Although Beijing denied foreknowledge of the North Korean missile test,[24] we suspect General Xiong gave the North Koreans the go signal. Just nine months before the test, General Xiong had assured Japan's Liberal Democratic Party leadership that North Korea was on the road to "peace and stability."[25]

The PRC's extensive involvement in North Korea leaves the Japanese concerned, and with good reason. Japanese military analysts believe the North Koreans have concluded that the August 31, 1998, missile test was so successful that they could begin deploying the missile in combat units. Five large underground launching sites are currently under construction, some near the Chinese border and some close to the Demilitarized Zone with South Korea. From there, North Korean missiles can reach all of Japan—and consequently all American forces stationed there—and may extend as far as northern Taiwan.[26]

By one estimate, North Korea has at least four missile factories, and perhaps as many as eight, all hard at work.[27] In 1998 U.S. intelligence reported that fifteen thousand laborers were building a vast underground nuclear facility that might be operational by the summer of 2000.[28] Months later U.S. inspectors found the facility empty, but the North Koreans may have moved the nuclear equipment before the inspectors arrived.

Our biggest concern is the possibility that the PLA might have transferred to North Korea some or all of the nuclear warhead designs the Chinese stole from the United States. North Korea would want the American designs not only because they are the most modern but also because they are light in weight. In a defensive maneuver, the North Koreans have moved their missile launch sites as far from Japan as possible, right up against the Chinese border,[29] because in 1999 the Japanese Self Defense Forces announced that a preemptive air strike against North Korean missile batteries would not violate Japan's constitution if Japan had reason to believe an attack is imminent.[30] Cutting down on the warhead's weight dramatically increases the range of a missile. So, with lightweight American warheads, North Korea could still threaten Japan with nuclear and biological annihilation.

The North Koreans would also want access to the neutron bomb technology General Xiong's spies stole from the Americans.[31] The neutron bomb attacks biological targets—that is, human beings—but causes minimal damage to structures. Like thieves in the night, the North Koreans would want to preserve Japan's valuable civilian infrastructure for their own use. The United States developed this technology but chose, wisely, not to deploy it. The PLA stole it, built it, and tested it, and now nothing is keeping the Chinese from selling it to the highest bidder.

Japan, like the United States, has no defense against ballistic missile attack.

MISSILES LOST, MISSILES FOUND

Shortly after the PLA navy was discovered to have taken over Philippine territory in the South China Sea, a senior Japanese diplomat in Washington invited a small group of American

national security specialists to dinner.[32] Approximately 60 percent to 70 percent of Japan's total energy imports pass very
close to Mischief Reef, he told them. Everyone present immediately understood the implications of the PLA's bold move:
Here was a not-so-subtle way for Beijing to control Tokyo's
lifeline.

What neither the Japanese nor the Americans knew at the
time was that Beijing was putting together a clever plan to back
up its forces in the South China Sea with a modern ballistic
missile system, the DF-25. A few years before, the DF-25 had
made a quiet appearance in an American scholarly journal.[33]
According to the authors of the article, the DF-25 is a mobile,
two-stage missile with a range of 1,100 miles—just enough to
reach the Spratly Islands from the Chinese mainland. The
authors reported that it had a massive two-ton conventional
high-explosive warhead, more than enough to destroy any oil
tanker.

But since it was thought to be non-nuclear, the DF-25 fell
off allied intelligence's radar screen. In 1996 the PLA told a
number of American, European, and Russian specialists that the
DF-25 had been canceled due to lack of funds.[34] That assertion
seems to have been accepted at face value, and it was not carried
on Western lists of Chinese missiles. Neither the Cox Report
nor *The Military Balance*, usually considered the definitive
resource, carried it.[35] Even the National Reconnaissance Office,
the American spy satellite agency, listed it as "under development," meaning its status was unknown.[36]

On August 3, 1999, allied intelligence received a nasty shock:
The DF-25 is very much alive, it is nuclear, it has multiple warheads, and it now comes in two versions—the 1,100-mile orig-
inal and a 1,500-mile "improved" version. Descriptions of the
DF-25 appeared simultaneously in two PRC-owned news out-
lets, a PRC-owned newspaper in Hong Kong, and a PRC-

owned press agency in Hong Kong.[37] There were no banner headlines, just small stories to let allied intelligence officials know they had been fooled completely.

The DF-25 poses an immediate threat to Japan. Its extended range and its two-ton warhead capacity make it possible to put a massive nuclear explosion anywhere in the country. Since the missile is solid fueled and mobile, the PLA can hide it in caves deep inside China and roll it out only when the Chinese Communists want to strike. This is a true "Theater" missile, able to threaten all the democracies in the region from Tokyo to Singapore. Certainly all of Japan's oil tankers are within range of the DF-25, whether armed with a conventional or nuclear warhead.

Allied intelligence and allied decision-makers now have a double problem. First, there's the threat the DF-25 poses by itself. Second, if the PLA fooled us on this one, how good are the rest of our estimates? The CIA estimates that about 23 long-range PLA missiles can now reach the United States mainland.[38] How do we know it's not 230? Or 2,300? It is our judgment that the United States may be ripe for a strategic surprise. The American intelligence community has gone politically correct, telling the Clinton-Gore administration only what it wants to hear about the PLA. Americans should not be surprised if the Japanese want their own spy satellite program.

THE FIRST INCURSIONS

Japan has for a long time controlled a number of small, uninhabited islands between the Japanese island of Okinawa and Taiwan.[39] But in recent years the PRC has claimed them. In the late 1990s Beijing has from time to time sent ships into the area, and at one point the PLA even sent in some of its new Russian Su-27 fighter aircraft.[40] Japanese Self-Defense Force flyers from Okinawa drove the PLA air force off, but the event was unsettling in Tokyo.[41]

Why the PLA would pick a fight with Tokyo is unclear. Some claim there is oil under the islands, but that is unlikely, given that there is no oil production within hundreds of miles. More significantly, these islands are the only place where Japanese and Chinese territory touch. The PRC has had

Thanks to Chinese technology, North Korean missiles can now target all of Japan—and the American armed forces who serve there.

border disputes with nearly all its neighbors, and this could be just one more encroachment. Another theory is slightly more sinister: The Chinese are using the incursions to probe Tokyo's resolve and test its defenses. So far the Japanese government has acted quickly and responsibly, but perhaps Beijing is looking for a day when Tokyo doesn't respond so vigorously.

THE THREAT OF INFORMATION WARFARE

In a June 1999 speech in Washington, Raisuke Miyawaki, a senior adviser to Japan's Commission on Critical Infrastructure Protection, told his American audience that, in essence, Japan is oblivious to the problem of information warfare and is only going through the motions of dealing with it in order to please the Americans.[42] "Japan's people and businesses have not yet fully realized that Japan is vulnerable to a cyberterrorist attack, and the effects such an attack would have on Japan," he said. According to Miyawaki, a "leadership void" in Tokyo at both the corporate and governmental levels exacerbates "a lack of technology understanding." Japan is only now doing something because "the rest of the world's biggest nations are pressuring Japan to focus on this issue."

Miyawaki is a cautious official with a distinguished record in and out of government, including time spent as a spokesman for Japan's prime minister. When the person who is essentially

Japan's information warfare czar publicly criticizes his nation's business and government circles, the problem must indeed be severe.[43]

Japan occupies territory about the size of California but has a population of 126 million, nearly half that of the United States. Further, Japan is very mountainous, so most of the population is packed into a limited number of valleys and plains. A successful information warfare attack would likely devastate hundreds of thousands, if not millions, of people.

Japan is a modern country with an economy second only to America's, and it is just as exposed to attacks against its critical infrastructures as the United States. As with the United States, Japan has many areas that could be subject to an information warfare attack: oil and gasoline refineries, defense-related industries, food-processing plants, water and sewer systems, telephone systems, banks, and insurance companies.

In some areas, however, the Japanese are even more vulnerable than Americans. For instance, a hostile information warfare campaign would target the Japan National Railways. Unlike in the United States, where freight would be the prime target, passengers would be the most vulnerable. Groups have already attacked the computer control systems for commuter trains in Japan, paralyzing major cities for hours.[44] In 1995 the American defense think tank RAND conducted an information war game in which a high-speed train traveling at 180 miles per hour slammed into a misrouted freight train in the state of Maryland.[45] RAND's specialists estimated sixty American deaths from such a wreck, but a surprise attack on the Japan National Railways would kill many more people. With multiple accidents at the height of rush hour followed by a total shutdown of the emergency response system, those who didn't perish immediately from the collisions would die lingering, painful deaths because rescue teams would not be able to reach the victims.

The power grid, especially the nuclear component, is another likely target. Japan generates 17 percent of its electricity from fifty-two nuclear power plants.[46] If Miyawaki is correct about Japanese business circles' being generally unprepared to defend against a clever and determined information warfare attack, the nuclear power plants may be vulnerable. The meltdown at Chernobyl in 1986 involved one plant located in a mostly rural area; no one has any idea the damage possible from multiple, simultaneous Chernobyls occurring in an urban environment like Japan.

OUTLOOK

By arming North Korea with long-range missiles, the PRC has brought a direct threat to Japan. Thanks to Chinese technology, North Korean missiles can now target all of Japan—and the American armed forces who serve there. The PRC may have even provided North Korea with the nuclear technology their spies stole from the United States.

What's more, it appears the Chinese have begun their own encroachments on Japan, following a pattern of territorial aggression that threatens all of the PRC's neighbors. As the PLA builds up its extensive information warfare capabilities, Japan remains woefully unprepared for an attack. In the coming century, the Japanese people may discover the danger that comes with unpreparedness.

CHAPTER 12
SURROUNDING INDIA

China is potential threat No. 1.... The potential threat from
China is greater than that from Pakistan and any person who is
concerned about India's security must agree with that fact.[1]

—Indian Defense Minister George Fernandes, 1998

In the summer of 1994 the Indian coast guard seized three
PRC "fishing" vessels in Indian waters off the Andaman
Islands. The boats, flying the Burmese flag, had no fish
aboard, but they did have military maps and charts as well as
sophisticated electronics gear. Not wanting to alarm the public, the Indian government tried to hush up the incident, but
word leaked out.[2]

What were the Chinese ships doing?

It is our military judgment that the Chinese trawlers were
caught mapping the floor of the Bay of Bengal and the Andaman Sea in preparation for the introduction of Chinese ballistic missile submarines (SSBNs), just as the Soviet Union mapped
the floor of the North Atlantic in the 1950s. *Before* the SSBNs
arrive off India's coast, the PLA navy needs to know every hid-
ing place, every crack and fissure in the ocean floor, which
requires years of meticulous preparation.[3]

The close relationship between the PRC and the SLORC enabled the PRC trawlers to home-port in Burma and fly the Burmese flag. The Burmese military junta has also allowed the PLA to build a missile-tracking station on Coco Island, a Burmese possession in the Indian Ocean near India's missile launch site. There are even unconfirmed reports that the SLORC is allowing the PLA to build military facilities all up and down the Burmese coast, facing the Andaman Sea and covering the entrance to the Straits of Malacca.[4]

Before **India set off its first nuclear explosion, the Chinese were establishing a nuclear weapons program for its bitter enemy, Pakistan.**

Some Burmese exiles have taken to calling their country (in private) the "Burmese Autonomous Region," evoking the name the PRC gives to Tibet ("Tibet Autonomous Region"). By complying with the PRC's creeping invasion of Burma, the SLORC has permitted the Chinese to further their aggressive designs on India.[5]

An Indian defense planner could be pardoned if the PRC's actions have raised suspicions, especially after Chinese Defense Minister Chi Haotian's November 1998 visit to Mozambique.[6] Chi is a very busy man, and under ordinary circumstances he wouldn't spend four days in a moderate-sized African country. But Mozambique has almost a thousand miles of coastline on the Indian Ocean. Chi's visit to Mozambique is particularly troubling in light of the inroads the PLA has systematically made into the nations surrounding India:

- Tibet, to the north of India, is the traditional buffer state. According to the Tibetans, the PLA's Second Artillery has nuclear-tipped missiles stationed in Tibet—missiles pointed at India's teeming cities and defense installations.
- In Bangladesh, which borders India to the northeast, the PLA sells arms and trains the military.[7]

- Burma, to the east, is fast becoming a PLA satellite.
- In the island nation of Sri Lanka, to the south, the PLA is entrenched as arms dealers and military trainers.[8]
- The island of Madagascar, in the Indian Ocean off of Mozambique, has seen an upsurge of PLA visits and offers of military training.[9]
- And bordering India to the northwest is Pakistan, India's bitter enemy.

THE PRC ARMS INDIA'S ENEMY

Sometimes persistence pays off. Mikio Haruna of Japan's Kyodo News Service has covered proliferation stories for more than twenty years.[10] In May 1989 he filed a Freedom of Information Act request with the CIA asking for "records concerning the development of nuclear capability in Pakistan."[11] In July 1995—six years after his initial request—he finally received an answer, containing portions of a 1975 U.S. Army intelligence report:

> Sometime before Oct. 74 the PRC assigned 12 scientists (including two experts and their assistants) to assist Pakistan in developing its nuclear science. The PRC is training nuclear science-related technicians for Pakistan, providing her with nuclear energy equipment and helping her build nuclear bases. One such base is located in Karachi; another is scheduled for construction in 1975. The latter reportedly will be used exclusively for nuclear weapons research.[12]

So, *before* democratic India set off its first nuclear explosion in October 1974, the Chinese were already establishing a nuclear weapons program for its neighbor (and enemy) Pakistan. Worse, American intelligence knew it and didn't tell the Indians.[13]

Haruna received a bonus package from U.S. Army intelligence—a report dated April 1983 with additional details of China's assistance to Pakistan's strategic weapons programs. According to the report, China was providing "considerable assistance" to Pakistan's "atomic bomb" program at a restricted "weapons research section" adjacent to the Kanupp nuclear power station in Karachi. Chinese and Pakistani military leaders drove the nuclear scientists without regard for human life: the U.S. Army report revealed the existence of a secret graveyard within the restricted area for at least six victims of radiation poisoning. And, for the first time, American intelligence reported that the Chinese were providing Pakistan with assistance in "the development of rockets and satellites."[14]

In February 1996, after the press began to reveal the extent of Chinese assistance to the Pakistani nuclear weapons program,[15] CIA Director John Deutch confirmed the PRC's illicit nuclear transfers. Deutch told the Senate Select Intelligence Committee that the Chinese had transferred five thousand samarium-cobalt ring magnets, identical to the ring magnets China had sold to Saddam Hussein's nuclear weapons program before the Gulf War.[16] According to Mark Hibbs of *Nucleonics Week*, with five thousand ring magnets Pakistan could increase its capability to enrich uranium by 100 percent.[17]

On May 10, 1996, the State Department was forced to admit the following:

We believe that in this instance Chinese entities transferred custom-made ring magnets to an unsafeguarded Pakistani facility engaged in uranium enrichment and undertook other cooperation with unsafeguarded Pakistani facilities.[18]

Senator Robert Bennett (R-Utah) was not satisfied with State's answers and pressed for details: Which "Chinese entities"?

What "other cooperation"? Reluctantly, the State Department identified the "Chinese entities" as the China Nuclear Energy Industry Corporation, a subsidiary of the China National Nuclear Corporation. The "unsafeguarded Pakistani facility" turned out to be the Khan Research Laboratories, named for the father of Pakistan's bomb, Dr. A. Q. Khan.

The State Department still did not want to explain what "other cooperation" Chinese companies had engaged in, but it did make a critical revelation: The United States had "concerns" over Chinese assistance "to both weapons development and production of unsafeguarded nuclear materials."[19] That is, the United States had information, undoubtedly from intelligence sources, that the Chinese were providing Pakistan with both the nuclear ingredients and the technology to turn them into a weapon, duplicating their earlier efforts for Saddam Hussein in Algeria. All American intelligence agencies—with the exception of the Bureau of Intelligence and Research for Clinton's State Department—agreed that Pakistani weapons engineers had finished developing nuclear warheads, thanks to the Chinese.[20]

The weaponization issue is crucial because the technological jump from simply having nuclear materials to having the know-how to turn them into a nuclear weapon is significant. Weaponization is what America's ultra-secret Manhattan Project[21] was all about. If, as the State Department has admitted, China has weaponized Pakistan's illicit nuclear program, it makes a powerful statement about Chinese intentions.

There are basically two ways to make nuclear weapons—through enriched uranium, where the ring magnets are crucial, and through plutonium reprocessing. It is likely that the "other cooperation" the State Department didn't want to discuss is plutonium reprocessing. China is building a clandestine heavy-water nuclear reactor near Kushab, in Pakistan's Punjab province.[22] Reprocessing spent fuel from that reactor into plutonium would contribute significantly to Pakistan's strategic weapons program, since plutonium is lighter than enriched uranium and more easily adaptable to missile warheads. Recent reports reveal that the Chinese are helping the Pakistanis with this reprocessing plant,[23] which will give Pakistan a second and more deadly nuclear capability.

In October 1996 the *Washington Times*'s Bill Gertz embarrassed the Clinton administration further by reporting, based on a top-secret CIA document, that the China Nuclear Energy Industry Corporation was trying, again in secret, to ship high-temperature furnaces and nuclear diagnostic equipment to an unsafeguarded nuclear facility in Pakistan.[24] According to Gary Milhollin, a well-known nuclear weapons specialist and director of the Wisconsin Project on Nuclear Arms Control, the furnaces could be used to cast uranium or plutonium for nuclear weapons.[25] Quoting the highly classified document, Gertz reported the CIA's belief that "senior level [Chinese] government approval was probably needed for this most recent assistance." According to the document, the CIA officers told their superiors of Chinese officials' efforts to deceive the United States with regard to this and "future" sales to Pakistan.[26]

The Chinese have also armed Pakistan with M-11 missiles, frustrating American officials who have tried to curb Chinese arms sales. In April 1991 Bill Gertz and the *Washington Post*'s R. Jeffrey Smith broke the story that China had sold its highly mobile, modern, nuclear-capable missiles to Pakistan,[27] which sounded alarm bells in Washington. Just one year before, India and Pakistan had nearly come to an all-out nuclear exchange, which, according to then–CIA Deputy Director Richard Kerr, "was the most dangerous nuclear situation we have faced since I have been in the U.S. Government."[28] During the 1990 confrontation Pakistani military leaders had searched unsuccessfully for a delivery system and had even contemplated using a C-130 transport plane to drop an atomic bomb on New Delhi.[29]

So Washington didn't want a solution to Islamabad's nuclear delivery problems, but as usual Beijing was not listening. The April 1991 M-11 delivery was composed only of Transporter Erector Launchers (TELs)[30] and practice (inert) missiles,[31] but the real thing was in the pipeline.

In July 1991 President Bush imposed sanctions on two Chinese companies identified in the M-11 transfer—Great Wall and China National Precision Machinery Import-Export Company. After a year-long diplomatic effort, on February 1, 1992, Secretary of State James Baker extracted a written pledge from Chinese Foreign Minister Qian Qichen that the PRC would live up to the guidelines of the Missile Technology Control Regime (MTCR)[32]—that is, that the Chinese would not transfer M-11 missiles or missile technology to Pakistan.[33] The cut-off pledge applied to the existing April 1991 sale; in other words, there was no "grandfather clause."[34] In May 1992 a Bush administration official stated that the administration was "100 percent sure" the missiles would not be shipped. "The deals are off," he said.[35] Bush lifted the sanctions.

Barely nine months after Beijing's pledge, however, U.S. intelligence caught Chinese military firms delivering M-11s to Pakistan.[36] Chinese arms dealers, aware of President-elect Bill Clinton's tough campaign stance against Bush's supposedly weak anti-proliferation record on China, evidently felt that they should make their deliveries before Clinton took office in January 1993. Citing U.S. government sources, the *Independent* of London reported, "There is clearly an explosion of deals in the making and the Chinese are clearly number one in the middle of it."[37]

But the Chinese didn't stop the transfers once Clinton took office. By May 1993 the *New York Times*, citing American intelligence reports, was reporting that Chinese firms had again broken the PRC government's pledge not to ship missiles to Pakistan.[38] The Russian Foreign Intelligence Service confirmed the proliferations in a 1993 report.

Under increasing bipartisan pressure from Congress, the Clinton administration in August 1993 sanctioned a number of Chinese companies, including Great Wall (again) and China National Precision Machinery Import-Export Company (again), for M-11–related transfers to Pakistan.[39]

Just as James Baker had done, Secretary of State Warren Christopher spent a year wrangling with the Chinese before Foreign Minister Qian Qichen signed a statement pledging that "China will not export ground-to-ground missiles featuring the primary parameters of the Missile Technology Control Regime."[40] So, the Chinese pledged (again) to live within the boundaries of the MTCR, and the American secretary of state (again) lifted the sanctions against Chinese companies.[41]

At a meeting with PRC President Jiang Zemin later in the year, President Clinton, nervous about the Christopher-Qian deal, offered to exercise his waiver authority for any *future* Chinese missile trade violations if the PRC would only reveal the extent of its existing missile trade with Pakistan.[42] Jiang turned him down—and with good reason, as the American intelligence community discovered nine months later. On July 3, 1995, the *Washington Post* reported that American intelligence agencies had "incontrovertible proof" that China had recently transferred M-11s to Pakistan.[43]

By early 1996 the intelligence community had learned that the M-11 transfers were just a small part of China's illicit arms sales. CIA Director John Deutch told the Senate:

The Intelligence Community continues to get accurate and timely information on Chinese activities that involve inappropriate weapons and military technology assistance to other countries: nuclear technology to Pakistan, M-11 missiles to Pakistan, cruise missiles to Iran.[44]

In April 1996 Hong Kong Customs agents intercepted two hundred boxes of Chinese ammonium perchlorate, a rocket fuel prohibited under MTCR; the fuel was destined for Pakistan's Space and Upper Atmosphere Research Commission, a missile research center. Hong Kong authorities charged the China Ocean Shipping Company (COSCO), the PRC government–owned shipping line that has a history as a delivery service for the PLA arms merchants.[45] In 1991 COSCO was discovered making midnight arms deliveries to the Burmese military regime on behalf of Poly.[46]

The People's Liberation Army's incursions into South Asia have left India a nation surrounded.

In June 1996, after a closed door briefing from Gordon Oehler, director of the CIA's Nonproliferation Center, Representative Floyd Spence (R-South Carolina), chairman of the House National Security Committee, and Representative Curt Weldon (R-Pennsylvania), chairman of the House National Security Subcommittee on Research, told the press that Chinese M-11s were indeed present in Pakistan.[47] The American intelligence agencies were nearly unanimous in stating that the M-11 missiles were not only present, but were already "operational." (Again, only the Bureau of Intelligence and Research for Clinton's State Department expressed reservations about this conclusion.)[48] In late summer 1996 the *Washington Post*, citing American intelligence officials, declared that Beijing was building an entire M-11 factory near Rawalpindi.[49] A year later

Pakistan announced the successful test flight of the Hataf-3, a missile with a range of five hundred miles probably based on Chinese technology.[50]

So much for the Chinese government's most solemn commitments to the United States. By the fall of 1996 Secretary of State Christopher could only complain to Foreign Minister Qian at their New York meeting about the need for "implementation" of agreements. Qian, however, treated this request with contempt.[51]

In short, the Chinese government stands accused of transferring the entire nuclear weapons cycle to Pakistan: technology to make the nuclear materials, the nuclear weaponization program, and the missile delivery system.

Looking back on his own unsuccessful attempt to halt Chinese missile sales to Pakistan, former Secretary of State James Baker wrote: "I suspected why [the effort had failed]: the Chinese had signed lucrative contracts to deliver missiles to Pakistan. In all probability, several senior government and party officials or their families stood to gain from the performance of those contracts."[52] Baker did not identify the "party officials or their families," but we know that it was Poly Group, headed by Deng Xiaoping's son-in-law.[53]

INDIA SURROUNDED

Although democratic India poses no threat to China, the PLA has targeted India as part of its long-term drive toward regional dominance. The PLA's incursions into South Asia have left India a nation surrounded. The PLA once directly attacked India and seized thousands of square miles of Indian territory. But now the Chinese seek to keep India's power in check by controlling the rest of the South Asian neighborhood.

Integral to the PRC's campaign against the Indians has been its continued weapons transfers to Pakistan, India's sworn enemy. In 1990 India and Pakistan came to the brink of nuclear confrontation. At that time, the Pakistanis did not have a delivery system for their nuclear weapons. Now they do—thanks to the Chinese illicit arms sales.

More disturbing may be that the Clinton-Gore administration is an accomplice to the PLA's designs. For almost seven years the administration has watched as PLA companies transfer weapons of mass destruction equipment and technology to Pakistan—and many other rogue states. The Clinton-Gore foreign policy, which places trade and commerce ahead of national security, blinds the administration to the rampant proliferation reported by the allied intelligence agencies and America's own press corps.

The Clinton-Gore administration has sent the message that the United States will tolerate weapons proliferation.

In September 1995 President Clinton's ambassador to India bluntly told a delegation of four senior congressional aides that the administration wanted to repeal the Pressler Amendment, which limited U.S. arms sales to Pakistan because of the Pakistani regime's nuclear weapons program.[54] The Clinton-Gore administration, said the ambassador, wanted to sell arms to *both* India and Pakistan.[55] The aides—two Democrats and two Republicans—were stunned: The White House wanted to sell modern weapons to the two enemies—both nuclear powers— that most foreign policy experts put at the top of their list of potential flashpoints. It was the Clinton-Gore mercantilist foreign policy in its rawest form.

This slavish adherence to a foreign policy of dollars before national defense and "strategic engagement" with the PRC has sent the message that the United States will tolerate weapons proliferation. In the spring of 1998 the Clinton administration's assistant secretary of state for South Asia declared that the administration's number one policy goal for

the region was trade and commerce; the Indians and Pakistanis answered with the largest number of nuclear weapons and
ballistic missile tests ever conducted in so short a time.

The PRC has learned this lesson all too well, as it has repeatedly violated its most solemn commitments not to transfer
weapons to rogue regimes—and has not suffered any consequences for its scofflaw proliferations.

In the coming years, the United States and its democratic
allies may learn the devastating effects of the Clinton-Gore
administration's misguided foreign policy.

CHAPTER 13
THE VIEW TOWARD THE FUTURE

No regime poses a greater threat to global security today than Communist China. With the collapse of the Soviet Union, and the internal disarray of the Russian Federation, the People's Republic of China sees itself as the sole Communist superpower in the world. With its strategic military buildup—aided by stolen and acquired American weapons technology—the Communist Chinese leadership has sought to make the world's most populous country one of its most powerful.

As we've seen, the PRC's territorial ambitions are immense. Now, armed with the most modern weapons of mass destruction and sophisticated instruments of information warfare, the Communist Chinese military threatens to take the lands it has historically coveted. We have already witnessed the PLA's recent incursions into Japanese territory as well as its strategic power projection in the Spratly Islands, territory of the Philippines. The island of Taiwan, just a hundred miles from the Chinese mainland, is in the PRC's sights; now that Beijing has the military might, the Communist regime wants to finish off the Chinese Civil War, half a century later.

Communist China sees but one major roadblock to achieving regional hegemony—the Pacific stretch of the United States. Consequently, the PRC's military buildup has been tailored to counter America's military capabilities—from its nuclear missiles, to the Sovremenny destroyers, to its information warfare program. America is now a target.

Faced with this threat, what has been the response of the Clinton-Gore administration? It has:

- permitted the sale of hundreds of supercomputers to the Communist Chinese, exposing America to the threat of destructive information warfare;
- failed to enforce American anti-proliferation laws, so that it is only a matter of time before some PLA weapon is used by a rogue state in the Middle East or South Asia, raising the spectre of germ or nuclear warfare;
- allowed PLA rockets to be made more reliable by the technical assistance of Loral, an American corporation whose president, Bernard Schwartz, was the 1996 Clinton-Gore campaign's leading donor;
- referred to Taiwan as "the turd in the punchbowl" of U.S.-China relations, essentially selling out a democratic government to a tyrant and making war in the Taiwan Straits much more likely;
- watched as a Communist Chinese company gained control of the Panama Canal;
- encouraged nuclear proliferation by refusing to crack down on North Korea or recognize that it is a surrogate for PLA ambitions, leaving countries like Japan to consider building their own nuclear weapons for self-defense;
- shown so little interest in our Pacific allies that the South China Sea might become a Chinese Communist lake; and

- exposed the United States to derision and distrust by the way Communist Chinese agents have so easily been allowed to buy influence at the White House.

As the Communist Chinese sold weapons of mass destruction to rogue states and terrorist regimes and used American military technology to enhance their own weapons systems, the Clinton-Gore administration turned a blind eye. The White House's failure to act has placed Americans directly in the line of fire.

Communist China sees but one major roadblock to achieving regional hegemony—the United States. America is now a target.

All the while, President Clinton hollowed America's military. Now, an overextended United States military—the only deterrent to Chinese aggressions—faces an increasingly dangerous People's Liberation Army.

By ignoring the PRC's fifty-year history of naked aggression and the threats to the United States and its democratic allies, Clinton and Gore have abetted Communist China's military ambitions. Perhaps most odious has been how the administration has rehabilitated the image of the PLA, a brutal army responsible for tens of millions of dead civilians in China, Korea, Tibet, and elsewhere. Under Clinton and Gore, six of the seven active-duty uniformed generals directly responsible for the June 1989 Tiananmen Square massacre have been welcomed to Washington.

It is a policy of appeasement that has failed to halt Communist China's military buildup and territorial aggression.

TO SLAY THE DRAGON

The future does not have to look like this—a future of war and Communist expansion abetted by Clinton-Gore corruption.

With honest, competent, and determined leadership, we can restrain, inhibit, and even perhaps de-fang Communist China as we shore up American security.

China policy isn't hard. Our greatest allies are in China itself—among the Chinese people. The spirit of the brave men and women who gave their lives at Tiananmen Square has not been extinguished, and rather than rolling out the red carpet for the butchers of Beijing, as the Clinton-Gore administration has done, we should offer constant encouragement to China's embattled democrats.

Moreover, we should, as far as we can:

■ suppress the PLA's modernization efforts—we should not be sharpening the sword that could be used against the Chinese people, our regional allies, or even ourselves;
■ stop PLA companies from selling missiles, poison gas, germ warfare equipment, and nuclear technology to the world's foulest tyrannies; and
■ recognize the Communist Chinese threat for what it is— the greatest national security danger to America as we open a new century.

And finally, the nub of the issue is this: We need to restore to the White House leadership that does not put America's foreign policy and security, and that of our allies, and the issue of human rights, on the auction block for campaign donations and commercial favors to cronies.[1]

America needs to put its own house in order by choosing a foreign policy based on honesty, principle, and a defense of freedom and democratic values—from brutalized Tibet, to the fanatical Communist regime of North Korea, to the blood-stained regime in Beijing itself. If we stand up for freedom, so too can China's patriots, and so too can all the peoples of Asia and the Pacific Rim.

As for us, our thoughts are never far from the dead patriots at Tiananmen Square. It is they who have been betrayed by the Clinton-Gore administration, and it is to them that the new, free China will look for inspiration when it bursts free from the murderous PLA.

NOTES

A NOTE ON SOURCES

Between us, we have more than fifty years of professional experience in national security affairs. One of us flew the F-4 Phantom II in Southeast Asia and developed an expertise in weapons systems and power projection. The other served as a China military analyst with the American intelligence community and spent decades as a China adviser to the Executive Office of the President and the United States Senate. Together we have an intense appreciation for the future strategic threats facing the United States and the rest of the democratic world.

The materials in this book are derived from dozens of interviews with eyewitnesses and experts here and abroad. A number of American intelligence documents make their first declassified appearances in these pages. Much of the information presented here is new—the emerging danger to the United States from cyberwarfare and the Clinton-Gore administration's secret program to rehabilitate the butchers of Beijing, to name just two items. Also, for the first time, the reader will learn of the PLA's clandestine effort to present Saddam Hussein with an offshore strategic weapons system that was stopped only by the good work and personal courage of the allied intelligence services. There are also the links between an arms smuggler with White House connections and the vile Burmese military junta, which sells heroin on America's streets to help pay for the arms dealer's wares.

In our first book, *Year of the Rat*, we documented how the Communist Chinese penetrated the American government through campaign contributions to Clinton and Gore. *Red Dragon Rising* shows how, under the President Clinton and Vice President Gore, the United States has ignored its role as leader of the democratic world, allowing the leading national security threat to grow ever more dangerous. In other words, they took the money and ran.

CHAPTER 1

1 "China, U.S. Issue Joint Statement," COMTEX Newswire, October 29, 1997, Xinhua.

2 That same night, just a few blocks away at the Washington Hotel, hundreds of guests gathered at another dinner to consider the real record of Communism in China—the brutal conquest of Tibet, the continued aggression against neighboring democratic countries, the spread of weapons of mass destruction to terrorist regimes, and, above all, the scores of millions who perished in China at the hands of the Chinese Communist Party and its military arm,

the People's Liberation Army. One of the authors, William Triplett, and his wife were in attendance that night, as were actor and civil rights activist Richard Gere, Representative Nancy Pelosi (D-California), and Representative Christopher Cox (R-California).

3   "Report of the Select Committee on U.S. National Security and Military/Commercial Concerns with the People's Republic of China," [henceforth "Cox Report"], Overview.

4   "The Select Committee judges that despite the 1998 announcement that the PRC and the U.S. would no longer target each other with nuclear weapons, the PRC's missiles remain targeted at the United States" (Cox Report, Vol. I, 183). We concur.

5   The six generals are: General Chi Haotian, in operational command at Tiananmen; General Xu Huizi, in tactical control; General Li Jijun, Deng Xiaoping's personal military aide; General Kui Fulin, the PLA operations director; General Xiong Guangkai, director of PLA intelligence; and General Zhang Wannian, in charge of paratroopers.

In most cases, the generals receive a nineteen-gun salute at the Pentagon, an Oval Office visit with President Clinton, and a visit to American military facilities we feel should be off-limits. For example, Chi visited the Los Alamos National Laboratory—from which Chinese spies stole American nuclear technology secrets—in civilian clothes so as not to attract attention. Typically, the controlled press in China spreads accounts of them across the front pages of the newspapers, with color photographs of the highest ranking American official possible.

6   See the authors' *Year of the Rat* (Regnery, 1998), 154–155.

CHAPTER 2

1   "The Death of a People's Army," by Brigadier General Michael T. Byrnes, in *The Broken Mirror: China After Tiananmen*, edited by George Hicks (Harlow, 1990), 141.

2   Dr. Mirsky's account is based on an interview conducted by the author (Triplett) on December 12, 1995, in Hong Kong.

3   "People's Republic of China: Preliminary Findings on Killings of Unarmed Civilians, Arbitrary Arrests, and Summary Executions Since June 3, 1989," Amnesty International, August 1989, 8.

4   *Observer* (London), June 4, 1989. For this story, Dr. Mirsky won the International Journalist of the Year award, Britain's equivalent of the Pulitzer.

5   This account of the parents is confirmed by the Amnesty International report (see page 22).

6   This account of medical personnel is confirmed by the Amnesty International report (see page 22).

7   Quoted by Chairman Christopher Smith, hearing of the House Subcommittee on International Operations and Human Rights, December 18, 1996.

8   By some estimates two million people were in the streets of Beijing on the night of June 3–4, 1989 (*Mandate of Heaven*, by Orville Schell [Simon & Schuster, 1994], 139).

9   *Tiananmen Square*, by Scott Simmie and Bob Nixon (Douglas & McIntyre, 1989), 180.

10 *Tiananmen: The Rape of Peking*, by Michael Fathers and Andrew Higgins (Doubleday, 1989), 112.

11 *Quelling the People*, by Timothy Brook (Oxford, 1992), 123.

12 Ibid, 113–114.

13 Simmie and Nixon, 182.

14 Interview with John Pomfret, January 4, 1997, in Washington, D.C.

15 Ibid.

16 Fathers and Higgins, 125–126.

17 Interview with Pomfret.

18 *Boston Globe*, June 4, 1989.

19 *Washington Times*, June 5, 1989.

20 *Financial Times*, June 5, 1989.

21 Brook, 137.

22 Interview by the author (Triplett), in Beijing, April 1993.

23 *New York Times*, June 4, 1989. For their reporting, Sheryl WuDunn and her husband, Nicholas Kristof, the *Times*'s Beijing bureau chief, shared a Pulitzer Prize in 1989.

24 *Ta Kung Pao*, June 4, 1989, as quoted in BBC Summary of World Broadcasts, Section: Part 3 The Far East; B. Internal Affairs; 2. China; FE/0474/B2/1.

25 *Wen Wei Po*, June 5, 1989, as quoted in BBC Summary of World Broadcasts, Section: Part 3 The Far East; B. Internal Affairs; 2. China; FE/0474/B2/1.

26 *Wen Wei Po*, June 6, 1989, as quoted in BBC Summary of World Broadcasts, Section: Part 3 The Far East; B. Internal Affairs; 2. China; FE/0476/B2/1.

27 Reuters, June 5, 1989. Others may know the fate of this reporter, but we do not.

28 Testimony before the House of Representatives Subcommittee on International Operations and Human Rights, December 18, 1996.

29 Ibid.

30 Ibid.

31 Brook, 142.

32 *Baltimore Sun*, June 5, 1989.

33 Ibid.

34 Brook, 160. The U.S. Army counts flamethrowers as a regular part of a PLA infantry unit's weaponry. See "China Battlefield Development Plan," Defense Intelligence Reference Document NGIC-1122-30CH-96, July 1996, xviii (classified "Secret," portions declassified).

35 Brook, 160.

36 Ibid, 161.

37 *Washington Times*, June 9, 1989.

38 Ibid.

39 *St. Louis Post-Dispatch*, June 5, 1989.

40 "Massacre in Beijing: The Events of 3–4 June, 1989, and Their Aftermath," a report prepared by the International League for Human Rights, August 4, 1989, at page 2 (reprinted in Senate Hearing 101-1125, beginning at page 429).

41 *Wen Wei Po*, June 6, 1989, as quoted in BBC Summary of World Broadcasts, Section: Part 3 The Far East; B. Internal Affairs; 2. China; FE/0476/B2/1.

42 Interview with Pomfret.

43 *Burying Mao*, by Richard Baum (Princeton, 1994), 162.

44 Brook, 155.

45 Amnesty International report, 27.

46 Reportedly the 27th Group Army was responsible for this (*Washington Post*, June 7, 1989).

47 *Washington Times*, June 9, 1989.

48 *Eastern Express* (Hong Kong), June 11, 1996.

49 Former Party Chief Hu Yaobang.

50 *Black Hands of Beijing*, George Black and Robin Munro (Wiley, 1993), 219.

51 Schell, 131.

52 Baum, 276.

53 Black and Munro, 61.

54 Han Dongfang.

55 Black and Munro, 224.

56 Schell, 103.

57 In Changsha, Jinan, Xian, Nanjing, Shanghai, and Hangzhou. *China's Search for Democracy*, edited by Suzanne Ogden et al. (M.E. Sharpe, Inc., 1992), 241.

58 Ibid.

59 Ibid, 242.

60 *Crisis at Tiananmen*, by Yi Mu and Mark Thompson (China Books and Periodicals, 1989), 180.

61 Ibid, photograph after page 86.

62 "Statement of Support for the Students by Some Editors and News Reporters of CCTV," May 14, 1989, reprinted in Ogden, 214.

63 Mu and Thompson, 55.

64 Li remained premier until the spring of 1998, when he became the leader of the National People's Congress, China's rubber-stamp parliament.

65 As one student activist put it in 1989, "For us, Li Peng was the incarnation of everything wrong with China."

66 Chen report in Mu and Thompson, 206.

67 Ogden, 119–120.

68 Black and Munro, 256.

69 Sygma photograph found after page 90 in Fathers and Higgins.

70 Telephone interview with a Chinese-American human rights activist present in China in 1989, January 14, 1997.

71 *China Wakes,* Nicholas Kristof and Sheryl WuDunn (Times Books, 1994), 81.

72 *Beijing Spring,* photographs by David and Peter Turnley, text by Melinda Liu (Stewart, Tabori, and Chang, 1989), 85. Another policeman was seen holding a sign that read, "The students will surely win."

73 Schell, 101.

74 "Guarding China's Domestic Front Line: The People's Armed Police and China's Stability," by Tai Ming Cheung, *China Quarterly*, Fall 1996, 527.

75 Ibid.

76 Ibid, 539.

77 Black and Munro, 239.

78 *Deng Xiaoping: My Father,* by Deng Maomao (Basic Books, 1995), 467.

79 Deng gave up the chairmanship of the Party's Central Military Commission in 1989 but didn't give up the state Central Military Commission chair until the spring of 1990.

80 *The Chinese High Command,* by William W. Whitson (Praeger, 1973), 286, Chart G.

81 General Wang died in 1993.

82 Schell, 67.

83 *Who's Who in China: Current Leaders* (Foreign Languages Press, 1989), 722.

84 The PRC did not permit Muslim citizens to make the hajj to Mecca until 1979. After the Tiananmen massacre, the number of pilgrims declined precipitously

because the government would not grant exit permits. See "Country Reports on Human Rights Practices for 1990," U.S. Department of State, 856.

85 *Massacre in Beijing*, edited by Donald Morrison (Warner Books, 1989), 33.

86 Statement by an American army general, June 1995, in the author's (Triplett's) presence.

87 Excerpts of Yang Shangkun's May 24, 1989, speech to military commanders are reprinted in Mu and Thompson, 183.

88 Ibid, 184.

89 The authors' understanding of Tiananmen was heavily influenced by private discussions with allied military officers present in Beijing during the spring of 1989.

90 "The PLA and Its Role Between April–June 1989," by Tai Ming Cheung in *China's Military: The PLA in 1990/1991*, edited by Richard H. Yang (Westview Press, 1991), 2.

91 Reuters, June 7, 1989.

92 Morrison, 200.

93 Telephone interview with Ambassador Lilley, January 27, 1997.

94 Simmie and Nixon, 37.

95 Kristof and WuDunn, 79.

96 Ibid.

97 Simmie and Nixon, 37.

98 Brook, 29; Morrison, 34.

99 Mu and Thompson, 67.

100 *The Military and Political Succession in China*, by Michael Swaine (RAND, 1992), 46.

101 Byrnes, 141.

102 *Countdown to Tiananmen: The View at the Top*, by Jiang Zhifeng (Democratic China Books, 1990), 319.

103 Morrison, 33.

104 Interview with a British journalist present at Tiananmen, November 18, 1996.

105 "The Death of a People's Army," 139.

106 "Directory of P.R.C. Military Personalities," Defense Liaison Office, U.S. Consulate General, Hong Kong, June 1990, ii. The PLA Group Armies around Beijing on the night of the massacre were as follows:

        Beijing Military Region—24th, 27th, 28th, 38th, 63rd, 65th
        Chengdu Military Region—14th
        Guangzhou Military Region—15th Airborne Army
        Jinan Military Region—20th, 26th, 54th, 67th

> Lanzhou Military Region—47th
> Nanjing Military Region—12th
> Shenyang Military Region—39th, 40th, 64th

107 Byrnes, 133.

108 Interview with British journalist, in Hong Kong, 1996.

109 Ibid.

110 Statements by thieves organizations posted in Xian and read over Beijing University radio (reproduced in Ogden, 303).

111 Brook, 109.

112 Ibid.

113 Kristof and WuDunn, 86.

114 Chinese military intelligence operatives are thought to have tried to offer weapons to the students so that Beijing could later claim they were armed. The students quickly saw through that provocation.

115 Cheung, "The PLA and Its Role Between April–June 1989," 5.

116 *Washington Times*, June 9, 1989. See also "The Losses in Tiananmen Square," by Colonel Harlan W. Jencks, *Air Force Magazine*, November 1989, 62.

117 Black and Munro, 239.

118 Ibid, 218.

119 See Morrison, 35; Mu and Thompson, 83.

120 Morrison, 35.

121 *Washington Times*, December 11, 1996. See also the Associated Press story reprinted in the *Washington Post* of the same day.

CHAPTER 3

1 "Deng's Pyrrhic Victory," *The New Republic*, October 2, 1989. Liu Binyan was China's most prominent investigative reporter. He is now at Princeton.

2 Observation of an American citizen, July 1989.

3 The Tibet revolt of April 1989 should be considered an attempt to expel an occupying army, not a popular revolt against a home-grown Communist regime.

4 This figure is approximate; some regimes such as Mozambique could be counted as Communist-dominated.

5 The issue of whether Tiananmen was a "rebellion" is very sensitive in China. In September 1989 Jiang Zemin said, "We definitely can say it was a counter-revolutionary rebellion" (*Wall Street Journal*, September 27, 1989). By contrast, the pro-democracy forces consistently denied that they were attempting to overthrow the system. BBC correspondent James Miles comments, "After all, what happened between April and June 1989 was in essence a rebellion

against communist rule" (*The Legacy of Tiananmen*, by James Miles [Michigan, 1997], 38). We agree. The goals of democracy and human rights could not have been achieved within a Communist system. Liberty and Communism are incompatible.

6   *China After Deng Xiaoping*, by Willy Wo-lap Lam, P.A. (Professional Consultants, Ltd., 1995), 196. Lam is the China editor of Hong Kong's *South China Morning Post* and a noted authority on modern China.

7   *Burying Mao*, by Richard Baum (Princeton, 1994), 291.

8   Lam, 261.

9   "Country Reports on Human Rights Practices for 1989," U.S. State Department, 802.

10   AFP, May 30, 1998.

11   *China Directory 1996* (Radiopress, Inc., 1996).

12   "Directory of P.R.C. Military Personalities," June 1991, iv.

13   "Directory of P.R.C. Military Personalities," October 1994, v.

14   The general was Wang Yufa. *South China Morning Post* (Hong Kong), May 20, 1999.

15   These numbers are based on the Central Committee elected by the 13th CCP Party Congress (1987) and the committee chosen by the 14th CCP Party Congress (1992).

16   Lam, 198.

17   Baum, 305.

18   Lam, 218.

19   Ibid, 227.

20   Ibid.

21   Comment by Congressional Research Service specialist Shirley Kan, June 6, 1997, Washington, D.C.

22   Lam, 153.

23   Baum, 298.

24   Lam, 156.

25   Ibid, 245.

26   Ibid, 156–157.

27   "Country Reports on Human Rights Practices for 1996," U.S. State Department.

28   "China Country Report on Human Rights Practices for 1998," dated February 26, 1999, can be found on the Internet at <http://www.state.gov/www/global/human_rights/1998_hrp_report/china>. It is very disturbing reading. This is the most up-to-date U.S. government accounting of human rights in China.

29 Ibid.

30 For their intrepid reporting, Ross and Schwartz won an Emmy and three other prestigious journalism awards.

31 Constitution of the People's Republic of China, Articles 2 and 3.

32 AFP, June 13, 1997.

33 Miles, 180.

34 Dr. Shambaugh teaches at George Washington University and is the former editor of the *China Quarterly*. This quote is from UPI, printed in the *Hong Kong Standard*, April 22, 1994.

35 National Ground Information Center document 1122-30CH-96, classified "Secret," portions declassified, 4.

36 Ibid, 57.

37 Much of this discussion of the People's Armed Police comes from Cheung, "Guarding China's Domestic Front Line."

38 General Yang Guoping's rapid advancement may have been due to his troops' "effectiveness" during the Tiananmen Square massacre. In 1989 he was an officer of the Jinan Military Region's Shandong Military District, from which a number of troop units went to Beijing. He is also the son of an early Communist Party official (*Ming Pao* [Hong Kong], May 1, 1996).

39 *Jane's Defense Weekly*, October 28, 1995.

40 Beijing is particularly secretive about the People's Armed Police, so there are no reliable figures for People's Armed Police troop strength. One recent count in a Hong Kong publication (*Kuang Chiao Ching*, February 16, 1999) was 800,000, but that may be too low.

41 Cheung, "Guarding China's Domestic Front Line," beginning at 525.

42 Ibid.

43 *Jane's Defense Weekly*, October 28, 1995; Cheung, "Guarding China's Domestic Front Line."

44 *Beijing Renmin Wujing Bao* (in Chinese), May 12, 1998. This newspaper is published by the Political Department of the People's Armed Police.

45 *Beijing Renmin Wujing Bao* (in Chinese), March 3, 1998.

46 "No More Tiananmens," by Colonel John F. Corbett, Jr., and Dennis J. Blasko, *China Strategic Review*, Spring 1998. Colonel Corbett serves as senior country director for China in the Office of the Secretary of Defense.

47 Xinhua, November 3, 1995.

48 AFP, February 4, 1999.

49 Associated Press, January 15, 1999.

50 *South China Morning Post*, November 8, 1995.

51   *Fang-wei Ta Tai-wan* (Taipei, in Chinese), November 1, 1995.

52   *Quotations from Chairman Mao Tse-tung* (Foreign Languages Press, 1966), 61.

CHAPTER 4

1   In December 1997 the International Commission of Jurists in Geneva, Switzerland, declared that Tibet is "under alien subjugation." Reuters, December 22, 1997.

2   The Communists are very sensitive about their conquest of Tibet, given the obvious parallels with Hitler's September 1, 1939, invasion of Poland. As part of their Big Lie program, statements such as "The peaceful liberation of Tibet is a major event in the PRC's history since its founding" (Xinhua, December 11, 1994) are common.

3   Some useful documentation on the 1950 invasion of Tibet is found in the *Tibetan Sourcebook*, by Ling Nai-min (Union Research Institute, 1964), long out of print but found in some specialized libraries.

4   Interview with Mr. Lodi Gyari, president of the International Campaign for Tibet, Washington, D.C., December 29, 1994.

5   *Freedom in Exile, the Autobiography of the Dalai Lama of Tibet* (Holder and Stoughton, 1990), 115.

6   Ibid, 121.

7   The estimate of deaths comes from a captured PLA document (Ibid, 162).

8   "Tibet Under Chinese Thumb," by former Senator Larry Pressler, *Christian Science Monitor*, October 19, 1993.

9   This account cannot begin to do justice to the horrors of Communist China's occupation of Tibet. The International Campaign for Tibet, Human Rights Watch, and Amnesty International all do excellent reporting on the issue on a more-or-less continuing basis. For further reading, we particularly recommend His Holiness's two autobiographies and *Ama Adhe: The Voice That Remembers*, by Adhe Tapontsang (Wisdom Publications, 1997), the story of one Tibetan woman's struggle for freedom. The U.S. State Department's annual human rights reports are also useful.

10   Recent information suggests there is a high mortality rate among Tibetan women in Chinese prisons. Most of the deaths are the result of torture. Information provided by the International Campaign for Tibet, May 3, 1999.

11   Estimate by the International Campaign for Tibet, May 3, 1999.

12   Statement by the second CCP National Congress, quoted in *The Status of Tibet*, by Michael C. van Walt van Praag (Westview Press, 1987), 88.

13   *Eastern Express* (Hong Kong), April 13, 1995.

14   *The Forgotten War*, by Clay Blair (Anchor Books, 1989), 53.

15   This division was decided at Yalta.

16   For example, most of the output from North Korea's big Hungnam explosives plant was shipped to China for the PLA. See *The Origins of the Korean War*, Vol. 2, by Bruce Cumings (Princeton, 1981), 358–359.

17   *China's Road to the Korean War*, by Chen Jian (Columbia, 1994), 106. Chen's information is based on PRC sources.

18   Ibid, 108.

19   Ibid, 110.

20   *This Kind of War*, by T. R. Fehrenbach (Macmillan, 1963), 17–18.

21   The heroism of UN forces has been the subject of a number of book-length accounts.

22   In the Korean War, Seoul suffered through four major campaigns, first when the North Koreans invaded in the summer of 1950, next when the UN threw them out in the fall, then when the PLA returned in the winter, and finally when the Chinese were finally ejected in March 1951. As might be expected, Korea's ancient capital was devastated by the fighting.

23   *China Crosses the Yalu: The Decision to Enter the Korean War*, by Alan Whiting (Stanford, 1960), is the leading proponent of this theory.

24   See "The Cold War in Asia" (Woodrow Wilson International Center for Scholars, 1995–1996) for Russian accounts of meetings and conversations between the Communist leadership. *China's Road to the Korean War*, by Chen Jian, is also excellent on the new information.

25   Entire PLA military units had to be stricken from the rolls because they ceased to exist. According to Pentagon corridor talk, Chinese Defense Minister Chi Haotian took a battalion of about six hundred men to battle in Korea and only three survived.

26   Chen, 222.

27   *India's China War*, by Neville Maxwell (Anchor Books, 1972), 423. Maxwell was a leftist who was not sympathetic to the Indian cause, but at least some of this book is straight reporting.

28   *Himalayan Blunder*, by Brigadier General J. P. Dalvi (Thacker and Company, 1969), 364.

29   Whitson, 95.

30   Maxwell, 455–456.

31   Ibid, 468. At this point it's hard to remember when India's voice and Indian attitudes were taken seriously in foreign capitals. It took a series of nuclear blasts in the spring of 1998 (actions not recommended by the authors) to get India back on the front pages of Western newspapers.

32 *Freedom in Exile*, 191. This view of the effect of the PLA victory on Nehru is confirmed by one of his closest colleagues. See *India and World Politics: Krishna Manon's View of the World*, by Michael Becher (Oxford, 1968), 7. Maxwell is in agreement (see page 471). Nehru himself commented that India had been "living in a fool's paradise of our own making" by failing to take Beijing's aggressive intentions seriously (*Freedom in Exile*).

33 By this time the PLA had already tested a ballistic missile with a nuclear warhead.

34 *The Sino-Soviet Crisis*, by Richard Wich (Harvard, 1980), 97.

35 Ibid, 102–103.

36 *Foreign Relations of the People's Republic of China*, by John W. Garver (Prentice Hall, 1993), 306. Professor Garver teaches at Georgia Tech.

37 See "The Chinese-Vietnamese Border War of 1979," by J. J. Haggerly, *Army Quarterly*, July 1979, 266.

38 See also "China's Punitive War on Vietnam: A Military Assessment by Harlan Jencks," *Asian Survey*, August 1979, 805.

39 See *Brother Enemy*, by Nayan Chanda (Harcourt Brace, 1986), 358, and *New York Times*, March 27, 1979.

40 Haggerly, 268.

41 Soviet aid was essentially limited to some minor airlifts of Vietnamese soldiers from Laos and the south (*Far Eastern Economic Review*, March 9, 1979). The authors recognize that Vietnam's disastrous war in Cambodia and the end of the Soviet Union, combined with Communist policies at home, contributed to Vietnam's relative decline in the region.

42 Vietnam's veteran frontline troops were in Cambodia.

43 Title of a speech given by General Lin Piao in 1965 justifying "wars of national liberation."

44 Garver, 144.

45 Private comment by a serving American military officer who is a specialist on the PLA, May 1999.

46 See *The Tiger and the Trojan Horse*, by Dennis Bloodworth (Times Books International, 1986), 24. "The Communists were taking their line from Peking...."

47 See *Forbes*, July 28, 1997, for an account of the Kuok family.

48 Garver, 171.

49 For a distinctly leftist view of U.S. efforts to influence Indonesian politics, see *Subversion as Foreign Policy*, by Audrey R. Kahin and George McT. Kahin (University of Washington Press, 1995).

50 Garver, 151; also, personal information of the author (Triplett). The authors are aware that the PRC's role in the events of 1965 is disputed by some specialists but believe that our account will ultimately be proved the correct one.

51 *Revolution and Chinese Foreign Policy*, by Peter Van Ness (University of California, 1970), 102; also, conversations of the author (Triplett) with Indonesian military officers over time.

52 The authors do not defend what happened in Indonesia after the coup attempt but point out that there have historically been more summary killings after successful Communist takeovers, such as in China, Cambodia, and Russia.

53 *The Rise and Fall of the Communist Party of Burma*, by Bertil Lintner (Cornell, 1990), 25.

54 Ibid, 45.

55 "The Straight Zigzag: The Road to Socialism in North Viet-Nam," by Bernard Fall in *Communist Strategies in Asia*, edited by A. Doak Barnett (Praeger, 1963), 201, n7.

56 See *China and the Arms Trade*, by Anne Gilks and Gerald Segal (St. Martin's Press, 1985), 33–35.

57 See *The Battle of Dienbienphu*, by Jules Roy (Harper and Row, 1963), 4.

58 Gilks and Segal, 49.

59 Chanda, 12.

60 *The Black Book of Communism* (in French), edited and published by Robert Laffont (1997), contains an entire chapter devoted to the massacres perpetrated by the Khmer Rouge.

61 See "PLA Not a Threat," a commentary by former Secretary of State Henry Kissinger in the *Washington Post*, April 27, 1999.

CHAPTER 5

1 This account of Undersecretary Bartholomew's trip to Beijing is based on the authors' interviews with participants over time.

2 Under U.S. law (Chapter 7 of the Arms Export Control Act), a foreign company found to be proliferating certain kinds of missiles is subject to sanctions.

3 Interviews with participants, 1999.

4 This information comes directly from her biography printed in the annual report of her current (as of 1998) employer, China Aerospace Industrial Holdings, Ltd. (CASIL).

5   Lieutenant Colonel Liu is discussed extensively in the authors' *Year of the Rat*. For a chapter-length discussion of Chinese intelligence see Chapter 10 of *Year of the Rat*; for the long course see *Chinese Intelligence Operations*, by Nicholas Eftimiades (Naval Institute Press, 1994).

6   See *Los Angeles Times*, April 4, 1999, and July 3, 1999. Ji was replaced by Major General Luo Yudong in August 1999 (*South China Morning Post*, August 24, 1999).

7   Almost every international trade deal, whether it's wheat or artillery shells, is financed by at least one bank. *New York Times*, September 13, 1987.

8   The General Staff Department, the General Political Department, the General Logistics Department, and the General Armaments Department.

9   China National Precision Machinery and China Great Wall are subsidiaries of China Aerospace, as is her current employer, CASIL.

10 COSCO is treated extensively in the authors' *Year of the Rat*. "The Clinton Administration has determined that additional information concerning COSCO that appears in the Select Committee's classified Final Report cannot be made public" (Cox Committee, Vol. I, 46).

11 *Washington Post*, April 9, 1999.

12 *South China Morning Post* (Hong Kong), March 26, 1999.

13 Mr. Kong Dan is vice chairman of China Everbright Ltd. Our information comes from a well-connected Western press source.

14 Named for the sixteen Chinese pictograms that proclaim it. See Defense Intelligence Agency publication DI-1921-60-98 (unclassified).

15 Ibid.

16 These arms sales are covered at length in the next two chapters.

17 The AFL-CIO's Food Services and Allied Trades division has the most complete and up-to-date list of PLA products sold in the United States. See the following website: <http://www.kickthepla.org>.

18 See the Cox Report, Vol. I, beginning at 27.

19 In the mid-1980s CSS-2 intermediate-range ballistic missiles were sold to Saudi Arabia for an estimated $2.5 billion; they cost approximately $500 million to produce.

20 *China Daily, Business Weekly*, January 18, 1998.

21 Jiang is the adopted son of a "revolutionary martyr" (*Ming Pao* [Hong Kong, in Chinese], July 3, 1995).

22 Li Peng, formerly the premier, is the adopted son of the late Premier Zhou Enlai (*Ming Pao* [Hong Kong, in Chinese], July 3, 1995).

23 *Hong Kong Hsin Pao* (in Chinese), January 13, 1999.

24 "The People's Liberation Army Is Becoming 'The Princes' Army,'" by Xia Jia, *Front Line*, April 1996. Hong Kong's *Apple Daily* (*Ping Kuo Jih Pao* in Chinese) has a useful chart attached to its article entitled "Princeling Influence Within PLA Said Growing" (December 9, 1996).

25 Ji is thought of as something of a lightweight and may have achieved his position through his father, the late Ji Pengfei, a former foreign minister.

26 Zhang Pin, son of former Defense Minister Zhang Aipin, was in the Er Bu's Foreign Affairs Bureau as of late 1996. See *Apple Daily*, December 9, 1996.

27 "China's Red Princes," by Andrew and Leslie Cockburn, *Vanity Fair*, October 1993, 108.

28 This reportedly happened during the PLA's 1979 and 1984 battles with Vietnam. The best equipment was sold (for hard currency) to Iran and Iraq. See "China's Red Princes."

29 *Washington Post*, December 20, 1996.

30 Interview with the author (Triplett), 1988.

31 Carrie is owned by the PLA's General Political Department and is a fiefdom of the late Marshall Yeh's family (*Institutional Investor*, May 1995).

32 Marshal Nie's family. At one point he was the head of it, and later his son-in-law took command. His daughter was also a high official of COSTIND. Although Marshal Nie has passed on and his daughter and son-in-law have retired, COSTIND is still dominated by officials they nurtured and promoted.

33 For an excellent account of the CCP's concern over this, see *The Legacy of Tiananmen*, by James Miles, 67.

34 Interview with the author (Triplett).

35 General Ji was the head of the PLA's Intelligence Department (see testimony of Johnny Chung, May 11, 1999). In recent years, it seems that the U.S. Embassy in Beijing has been clueless about whom it is granting visas.

36 AFP (Hong Kong), August 11, 1999.

37 *South China Morning Post*, July 5, 1995.

38 *Newsweek*, November 18, 1991.

39 *China's Political System*, by June Teufel Dreyer (Allyn and Bacon, 1996), second edition, 153.

40 Conversation between the author (Triplett) and the deputy chief of mission.

41 *Jane's Intelligence Review*, March 1998, 36.

42 The junta now calls itself the State Peace and Development Council. For this account we are using the more well-known SLORC.

43 Aung San Suu Kyi is the mother of two grown sons. When her husband lay dying in the spring of 1999, the SLORC offered her an exit permit but not the

right of return. She chose to remain with her people, a sacrifice of great personal pain. The authors do not hide their admiration for her.

44   There seems to be no abomination to which the SLORC will not descend. It
has been accused of enslaving its own citizens as forced laborers on an international gas pipeline, assaulting minority peoples with germ warfare, and
even selling other minority children to China to be organ donors for the rich.

45   *Far Eastern Economic Review* (Hong Kong), October 3, 1996.

46   "Myanmar's Chinese Connection," by Bertil Lintner, *Jane's International
Defense Review*, November 1994, 23; also, *Jane's Defense Weekly*, August 20,
1993, 1. For Chinese assistance to Burma's SIGINT capability see *Jane's Intelligence Review*, March 1998.

47   *Jane's Intelligence Review*, November 1995, 515. *The Military Balance
1998/1999*, by the International Institute for Strategic Studies (Oxford,
1999), uses 1998 data to estimate the active-duty military in Burma at
434,800.

48   Commentary by a U.S. government official in *The Nation* (Bangkok),
March 21, 1997.

49   See "Burma's Economic Performance Under Military Rule," by Professor
Mya Maung, *Asian Survey*, June 1997, 503.

50   "Burma, the Country That Won't Kick the Habit," Anthony Davis and Bruce
Hawke, *Jane's Intelligence Review*, March 1998, 27.

51   *Towards Democracy in Burma* (Institute for Asian Democracy, 1992), 5. The
Institute for Asian Democracy is a small but effective human rights think tank
that the authors support financially. Senator Simon's $1 billion arms figure
was only through 1992. We estimate it has doubled since then.

52   Ibid, 22.

53   The Politburo member is Vice Premier Jiang Chunyun. See *Ping Kuo Jih Pao*
(Hong Kong, in Chinese), October 23, 1997.

54   This is based on the author's (Triplett's) conversation with a knowledgeable
Southeast Asian observer whose relative allegedly witnessed the transfer.

55   Equipment and technology to produce nuclear, chemical, or biological
weapons.

CHAPTER 6

1   Statement on the occasion of the U.S.–North Korea Nuclear Agreement,
October 14, 1994 (*Public Papers of the Presidents*).

2   "Algeria's Nuclear Ambitions," by Vipin Gupta, *International Defense
Review*, April 1, 1992, 329.

3   Ibid.

4   *Sunday Times* (London), April 28, 1991.

5   *Houston Chronicle*, November 11, 1991.

6   There are three basic elements to a nuclear weapon—the fissile or nuclear materials, a triggering mechanism to force the explosion, and a delivery system such as a bomb or missile.

7   *Sunday Times* (London), January 5, 1992.

8   *Washington Times*, April 11, 1991.

9   *Sunday Times* (London), April 28, 1991.

10 *Washington Times*, April 11, 1991.

11 *European*, November 15, 1991.

12 Senator Joseph Biden, floor speech, U.S. Senate, April 16, 1991.

13 Republic of Algeria Radio, Algiers (in Arabic), 1830 GMT, April 28, 1991 (BBC—ME/1059/A/1; April 30, 1991).

14 Chinese Foreign Ministry spokesman quoted by Xinhua, April 30, 1991. See also Reuters, April 30, 1991, noting the statement of officials of the Algerian Ministry of Scientific Research, as quoted in the "pro-government daily" *El-Moujahid*.

15 Algeria has a population of 28 million and armed forces of more than 120,000, while Libya, a country of 5 million, has armed forces of only 65,000 (*The Military Balance 1998/1999*, 123, 134.)

16 *The World Factbook 1995*, Central Intelligence Agency, 6.

17 *Far Eastern Economic Review*, May 8, 1991.

18 *Critical Mass*, by William Burrows and Robert Windrem (Simon & Schuster, 1994), 40.

19 "Iraq Inspections-lessons Learned," by Dr. Kathleen C. Bailey, United States Defense Nuclear Agency, Report Number DNA-TR-92-115 (declassified), January 1993, 75.

20 United States Army Intelligence and Security Command document dated May 12, 1986 (redacted version released under the Freedom of Information Act to the Nuclear Control Institute, Washington, D.C.).

21 Bailey, "Iraq Inspections-lessons Learned," 76.

22 "Conventional Arms Transfers to the Third World, 1983–1990," Congressional Research Service, 1991, Tables 2A, 2H, and 2I.

23 *Al-Ahram* (Cairo), April 22, 1991.

24 *Time*, December 16, 1991.

25 *Al-Ahram* (Cairo), July 22, 1991.

26 *Washington Times*, March 30, 1990; see also *New York Times*, April 24, 1990.

27 *The World Factbook 1995*, 274.

28 Ibid, 275.

29 *New York Times*, April 24, 1990.

30 We do not claim that all proliferation problems can be laid at the PRC's doorstep, only that the Chinese are the leading troublemakers in this area.

31 *Nucleonics Week*, May 7, 1991.

32 *Mideast Markets (Financial Times)*, December 11, 1989.

33 Letter of July 8, 1991, from the Defense Intelligence Agency to Mr. Paul Levanthal, Nuclear Control Institute, Washington, D.C.

34 AFP, October 3, 1991.

35 *Independent* (London), October 7, 1990.

36 *Independent* (London), September 30, 1990.

37 *Christian Science Monitor*, January 31, 1991.

38 *Observer* (London), April 28, 1991.

39 *Los Angeles Times*, May 18, 1991.

40 *Der Spiegel* (Hamburg), September 26, 1994; also, *Export Control News*, December 30, 1994.

41 Telephone interview with a United Nations Special Commission (UNSCOM) consultant who wishes to remain anonymous, October 1, 1996. See also *Strategic Digest*, April 1995.

42 South African Press Association news agency, Johannesburg, 1706 gmt November 22, 1994 (BBC Summary of World Broadcasts, AL/2161/A, November 24, 1994).

43 *Sunday Times* (London), March 28, 1993.

44 "Iraq and Weapons of Mass Destruction," by Dr. Anthony Cordesman, *Congressional Record*, April 8, 1992, S5063.

45 Burrows and Windrem, 198.

46 *The Death Lobby*, by Kenneth R. Timmerman (Houghton Mifflin, 1991), 256, 290.

47 *El Pais* (Madrid, in Spanish), August 23, 1998.

48 "Strategic Exposure: Proliferation Around the Mediterranean," by Ian O. Lesser and Ashley J. Tellis (RAND, October 1996).

CHAPTER 7

1 Secretary Cohen's answer to a question from Senator Edward Kennedy (D-Massachusetts) during the secretary's confirmation hearing before the Senate Armed Services Committee, January 22, 1997.

2 "The Acquisition of Technology Relating to Weapons of Mass Destruction and Advanced Conventional Munitions," a report by the Central Intelligence Agency to Congress, June 1997, 5.

3 *Washington Post*, January 19, 1998, quoting Beijing's *China Daily* of January 18, 1998.

4 *New York Times*, November 18, 1990.

5 Telephone interview with David Kay, former UN chief inspector in Iraq, July 1996.

6 From classified sources, the authors are aware of a number of PLA arms transactions that have never been made public.

7 *Washington Times*, February 19, 1997.

8 Under American law, the secretary of state determines that a country is ruled by a terrorist regime if "[t]he government of such country has repeatedly provided support for acts of international terrorism," Section 6(j) of Public Law 96-72.

9 This arms show, known as IDEX, is the world's largest and occurs in the spring of odd-numbered years.

10 This story was recounted by the vice president of a NATO-based aerospace company to the author (Triplett) within minutes of the incident's happening.

11 "Worldwide Maritime Challenges 1997," Office of Naval Intelligence, Washington, D.C., March 1997, 22.

12 *Qol Yisra'el* (Jerusalem), May 16, 1997.

13 *Mena* (Cairo), May 23, 1997.

14 *Washington Post*, October 30, 1991.

15 "The rumours are fabricated by arrogant Western powers and are not true" (statement by Iranian parliamentary speaker Mahdi Karrubi at a Beijing news conference, Associated Press, December 20, 1991).

16 The United States Central Command has responsibility for the Middle East.

17 Reuters, June 26, 1997.

18 Congressional Research Service Issue Brief IB92056, dated September 17, 1996, 4.

19 Letter of August 27, 1993, from Admiral William O. Studeman, acting director of the Central Intelligence Agency, to Senator John Glenn (D-Ohio), chairman of the Senate Committee on Governmental Affairs, printed in Senate Hearing 103-208, 185.

20 *Nucleonics Week*, May 2, 1991.

21 "An Assessment of Iran's Nuclear Facilities," by Greg J. Gerardi and Maryam Aharinejad, *The Nonproliferation Review*, Spring–Summer 1995.

22 Reuters, January 21, 1990.

23 Gerardi and Aharinejad, 210.

24 *Washington Times*, May 8, 1995.

25 *China Nuclear Industry News* (Beijing), November 20, 1996.

26 *Mednews*, July 22, 1991.

27   "The IAEA: How Can It Be Strengthened?" by David Kay, *Nuclear Prolifer-ation After the Cold War*, edited by Mitchell Reiss and Robert Litwak (Woodrow Wilson International Center for Scholars, 1994), 316–319.

28   *Washington Post*, March 13, 1998.

29   *Washington Times*, April 15, 1999.

30   Question submitted for the record to the Senate Foreign Relations Commit-tee, January 8, 1997; see also *Deseret News*, January 24, 1997, and *Wash-ington Times*, January 24, 1997.

31   Testimony of CIA Director John Deutch before the Senate Select Committee on Intelligence, Senate Hearing 104-510, 82.

32   "Cruise Missiles: The Discriminating Weapon of Choice?" by Amy Truesdell, *Jane's Intelligence Review*, February 1997.

33   Testimony of CIA Director John Deutch before the Senate Select Committee on Intelligence, Senate Hearing 104-510, 82.

34   Ibid.

35   Reuters, February 25, 1993.

36   Letter from Defense Intelligence Agency Director Clapper to the Senate Select Committee on Intelligence, January 25, 1994.

37   *Welt am Sonntag* (Hamburg), April 27, 1997.

38   Hearing of the House International Relations Committee, November 9, 1995, 14.

39   *Washington Post*, March 8, 1996.

40   Ibid.

41   Telephone conversation with R. Jeffrey Smith, August 22, 1996.

42   *Iran Brief*, May 5, 1997.

43   *Times* (London), April 6, 1997.

44   *Washington Post*, May 24, 1997.

45   Senator Frank Murkowski, floor speech, U.S. Senate, September 17, 1986, *Congressional Record*, S23689.

46   *Jane's Intelligence Review*, April 1992; see also *International Defense Review*, April 1994.

47   *Jane's Strategic Weapons Systems*, September 1995.

48   Ibid.

49   *Jane's Defense Weekly*, February 1, 1992.

50   *Defense News*, June 19–25, 1995.

51   "Ballistic Missile Shadow Lengthens," by Wyn Bowen, Tim McCarthy, and Holly Porteous, *International Defense Review*, February 1, 1997.

52   Ibid.

53   *New York Times*, June 22, 1995.

54   *La Stampa* (Turin), June 28, 1997.

55   *Jane's Defense Weekly*, May 1, 1996.

56   Associated Press, July 25, 1998.

57   Associated Press, August 2, 1998.

58   *Washington Times*, July 23, 1998.

59   *Institutional Investor*, July 1996.

60   *Newsweek*, July 4, 1988.

61   "Worldwide Threat to U.S. Navy and Marine Forces," ONI-2660S-007-93, Vol. II, CH-32.

62   *Washington Times*, February 10, 1996.

63   Ibid; also, *Washington Times*, March 27, 1996; also, Reuters, July 15, 1996.

64   Answer to question from Senator Arlen Specter (R-Pennsylvania), chairman of the Senate Select Committee on Intelligence, February 22, 1996, Senate Hearing 104-510, 13.

65   Answer to question from Representative Chris Smith (R-New Jersey), hearing ("Review of the Clinton Administration Nonproliferation Policy") before the House International Relations Committee, June 19, 1996 (printed in hearing record at page 8).

66   *Washington Post*, May 31, 1997.

67   This is based on unpublished but unclassified materials from the Office of Naval Intelligence.

68   Press briefing by Secretary Cohen at Manama, Bahrain, June 17, 1997.

69   Associated Press, June 17, 1997.

70   China National Precision Machinery Import-Export Corporation brochures in the possession of the author (Triplett).

71   *Jane's Strategic Weapons*, September 1995.

72   *Aviation Week*, October 30, 1995.

73   Xinhua, February 8, 1996.

74   UPI, January 30, 1993.

75   *Flight International*, December 13, 1995.

76   *Jane's Defense Weekly*, September 11, 1996.

77   According to Dr. Joshua Sinai, "Libya is on the verge of succeeding in developing a weapons of mass destruction (WMD) capability in the form chemical and biological weaponry (CBW) and the ballistic missiles to deliver them." Dr. Sinai is an executive with the Science Applications International Corporation, a company engaged in sensitive research for the American government ("Ghadaffi's Libya: The Patient Proliferator," by Dr. Joshua Sinai, *Jane's Intelligence Review*, December 1998).

78   Apparently, even the PLA had standards—at least at one point. *New York*

Times, April 16, 1991; see also Tel Aviv Educational Television Network, July 9, 1991.

79 *Washington Post*, April 28, 1992.

80 *Washington Times*, May 21, 1990.

81 Sinai, "Ghadaffi's Libya: The Patient Proliferator."

82 *New York Times*, June 7, 1990; also, *Washington Post*, June 7, 1990.

83 Associated Press, February 24, 1997.

84 Author's (Triplett's) telephone conversation with a State Department officer who wishes not to be identified, April 29, 1997.

85 *Washington Times*, July 12, 1990; see also Sinai, "Ghadaffi's Libya: The Patient Proliferator."

86 *Der Spiegel*, April 14, 1996; see also Sinai, "Ghadaffi's Libya: The Patient Proliferator."

87 "Chinese Arms Production and Sales to the Third World," by Richard Bitzinger (RAND, 1991), 13.

88 *Jane's Intelligence Review*, May 1996.

89 Ibid.

90 *Jerusalem Post*, December 12, 1989.

91 *Washington Post*, February 22, 1992.

92 *Washington Times*, June 18, 1998.

93 *Jewish Chronicle* (London), March 18, 1994.

94 Statement to the defense committee of the House of Commons, June 26, 1995 (reported by the *Financial Times* (London), June 27, 1995).

95 *Times* (London), April 1, 1995.

96 Sinai, "Ghadaffi's Libya: The Patient Proliferator."

97 *Washington Post*, February 11, 1992.

98 *Observer* (London), October 29, 1995.

99 *Los Angeles Times*, June 23, 1988.

100 *Al-Ittihad* (Abu Dhabi), July 31, 1989.

101 See Reuters, November 30, 1989.

102 *Los Angeles Times*, October 24, 1991.

103 *Asian Defense Journal*, August 1991.

104 *Forward*, September 27, 1991.

105 *New York Times*, March 5, 1992.

106 Ibid.

107 *U.S. News and World Report*, July 22, 1991.

108 *Ha'aretz* (Tel Aviv), May 26, 1995.

109 Testimony of CIA Director Robert Gates before the Senate Governmental Affairs Committee, January 15, 1992.

110 *Asian Defense Journal,* March 1992; also, *Newsweek,* June 22, 1992.

111 *Washington Times,* July 23, 1996, July 24, 1996.

112 ABC News, August 23, 1999.

113 This account is based on a series of conversations with Ambassador Bradley Gordon over several years.

114 "Threat Control Through Arms Control, Annual Report to Congress 1996," U.S. Arms Control and Disarmament Agency, dated August 13, 1997, 87.

115 "Proliferation: Threat and Response," Office of the Secretary of Defense, April 1996, 9.

116 Ibid.

117 *Toronto Sun,* June 20, 1999.

118 The following are China's primary obligations under the Nonproliferation Treaty:

ARTICLE I
Each nuclear-weapons state Party to the Treaty undertakes... not in any way to assist, encourage, or induce any non-nuclear weapons State to manufacture or otherwise acquire nuclear weapons or other nuclear explosive devices, or control over such weapons or explosive devices.

ARTICLE III 2
Each State party to the Treaty undertakes not to provide (a) source or special fissionable material, or (b) equipment or material especially designed or prepared for processing, use, or production of special fissionable material, to any non-nuclear weapons State for peaceful purposes, unless the source or special fissionable material shall be subject to the safeguards required by this article.

119 Answer to a question from Representative Chris Smith (R-New Jersey) on June 19, 1996, printed in "Review of the Clinton Administration Nonproliferation Policy" hearing before the House Committee on International Relations, 15.

120 "Threat Control Through Arms Control, Annual Report to Congress 1996," 71.

121 Article I of the Chemical Weapons Convention (officially known as the "Convention on the Prohibition of the Development, Production, Stockpiling, and Use of Chemical Weapons and on Their Destruction").

122 *Basic Documents in International Law,* edited by Ian Brownlee (Clarendon Press, 1995), 396.

123 MTCR ACDA Fact Sheet, July 16, 1996.

124 See the Arms Export Control Act.

125 See the Iran-Iraq Arms Non-Proliferation Act of 1992.

126 "The danger of proliferation is the most overriding security interest of our time," Secretary of State Madeleine Albright, speech to a NATO foreign ministers meeting, Brussels, December 16, 1997 (*Washington Post*, December 17, 1997).

127 "Every American should understand that weapons of mass destruction (WMD)—nuclear, biological, and chemical weapons and their means of delivery—pose a grave threat to the United States and to our military forces and to our vital interests abroad," Deutch-Specter Commission, July 1999. The commission is named after former CIA Director John Deutch and Senator Arlen Specter (R-Pennsylvania) and is officially known as the "Commission to Assess the Organization of the Federal Government to Combat the Proliferation of Weapons of Mass Destruction."

128 For a list of PLA-company arms sales through the first half of 1998, see the authors' *Year of the Rat*, 141.

129 In both cases the administration acted only after Congress became involved.

130 Personal experience of the authors.

CHAPTER 8

1 *Hsin Pao* (Hong Kong, in Chinese), February 9, 1996.

2 *Los Angeles Times*, June 18, 1999.

3 Russian specialists have raised this concern. See *Komsomolskaya Pravda* (Moscow, in Russian), December 16, 1998.

4 Someone has already done this, by breaking in through a vendor's computer system that was integrated into a baby-food manufacturer's network. Fortunately, the manufacturer found out in time. *Christian Science Monitor*, June 24, 1999.

5 As governments have desperately tried to keep costs down, they have gone more and more to computer-controlled prison gates and jail cell doors.

6 Modern gasoline refineries are controlled by a very few employees' opening and shutting valves by computer. Russian specialists have also raised this possibility. See *Komsomolskaya Pravda*, December 16, 1998.

7 In 1995 a Russian biochemistry student broke into a major New York bank's computer and transferred $12 million overseas. *Parameters*, Winter 1996–1997.

8 A frightening amount of information relating to a person's identity is avail-

able on the Internet. Armed with the various account numbers, a hostile force should be able to wreak havoc inside a major computer system.

9   American defense industry experts place information warfare and the technology associated with fighting this war in the sensitive category that stealth technology held in the late 1970s. The Department of Defense's "Joint Doctrine for Information Operations" (Joint Publication 3-13) is unclassified *except* for Annex I, which contains the information on Computer Network Attack (CNA).

10   For example, see *The Changing Role of Information Warfare*, edited by Zalmay M. Khalilzad and John P. White (RAND, 1999), 275.

11   Statement by Admiral John M. McConnell, director of the National Security Agency, 1995, quoted in *War by Other Means*, by John Fialka (Norton, 1997).

12   See, generally, the report of the President's Commission on Critical Infrastructure Protection, 1997.

13   Opinion of Dr. Michael Pillsbury, a renowned PLA expert at the U.S. National Defense University (Reuters, June 24, 1998).

14   From 1978 to 1981 one of the authors (Timperlake) worked on a project sponsored by Net Assessment and the Office of Strategic Research of the Central Intelligence Agency to make meaningful comparative assessments of military forces, focusing on the Soviet Union, the Warsaw Pact nations, and their Middle Eastern client-states.

15   *Guoji Hangkong* (Beijing, in Chinese), March 5, 1995.

16   The Tofflers are the authors of a number of books on the future of warfare. This particular quote comes from *War and Anti-War: Survival in the Dawn of the 21st Century*.

17   United States Army Field Manual 100-6, "Information Warfare."

18   At the moment the United States Joint Chiefs prefer to call it "Information Operations." See "Joint Doctrine for Information Operations," Joint Publication 3-13, October 9, 1998.

19   Ibid. Computer Network Attack (CNA) is the heart of American information warfare; even the name was classified until fairly recently.

20   Colonel Wang Baocun of the Academy of Military Sciences is one of the PLA's leading authorities on information warfare. His visionary piece appeared in *China Military Science* (Beijing, in Chinese), November 20, 1997.

21   Ibid.

22   Ibid.

23   Report to Congress Pursuant to Section 1226 of the Fiscal Year 1998 National Defense Authorization act.

24   Dr. Pillsbury is with the National Defense University. Dr. Mulvenon's statement was quoted in *The People's Liberation Army in the Information Age* (RAND, 1999).

25   "U.S.-Chinese Military Relations in the 21st Century," a paper prepared for the 1998 conference on the People's Liberation Army, Wye Plantation, September 11–12, 1998.

26   One of these scientists, Qian Xuesen, was an American missile pioneer who renounced his citizenship and returned to China in the 1950s. The Cox Report claims he was a spy for the Communists before he left the United States.

27   *Jiefangjun Bao* (Beijing, in Chinese), March 30, 1998.

28   *Sun Tzu, The Art of War*, translated by Samuel B. Griffith (Oxford, 1963).

29   *Jane's Intelligence Review*, August 1995. Professor Ball is reputed to have close ties to allied intelligence.

30   See the authors' *Year of the Rat*, Chapter 12. We believe the PLA is on the verge of a major expansion into space. See, generally, the Cox Report, Vol. II.

31   Most of this information comes from an account in *El Nuevo Herald* (Miami, in Spanish), June 24, 1999.

32   We recognize that if COSCO should use its U.S. facilitates for interception, American specialists would know fairly quickly by the shape of the antennas. But we also believe that the Clinton-Gore administration does not have the will to put a stop to such practices.

33   Cox Report, Vol. I, 144.

34   Ibid, 101.

35   Ibid, 105.

36   Ibid, 116.

37   Ibid, 138. Sun's Hong Kong affiliate was fined $10,000 (Dow Jones News Wire, June 21, 1999).

38   Cox Report, Vol. I, 41.

39   Ibid.

40   Xinhua (domestic news service, in Chinese), October 5, 1994.

41   *Jiefangjun Bao* (Beijing, in Chinese), December 25, 1995.

42   *Guoji Hangkong* (Beijing, in Chinese), March 5, 1995.

43   Website: <http://jya.com/rfw-jec.htm>.

44   Dr. Lin Weigan is not unique. The PLA's most significant strategic weapons programs owe their start to Chinese who were trained in the United States, renounced their American citizenship, and returned to China. For example,

see the account of Dr. (and PLA General) Qian Xuesen in *Thread of the Silk-worm*, by Iris Chang (Basic Books, 1995).

45   There's a lot of PLA open-source literature on high-powered microwave weapons from the early 1990s (for example, see *Dianzi Keiji Xuebao* [in Chinese], February and June, 1992) but somewhat less lately, perhaps indicating their systems have matured.

46   In the words of the Cox Committee, "During the late 1990s, U.S. research and development work on electromagnetic weapons technology has been illegally obtained by the PRC as a result of successful espionage directed against the United States" (Cox Report, Vol. I, xiii).

47   Apparently the Russians sold suitcase-sized high-powered microwave "bombs" to at least Sweden and Australia. See *Sevenska Dagbladt* (Stockholm, in Swedish), January 21, 1998.

48   It's unlikely that Jiang expected his words to be written down and circulated by the Party Central Committee, with a copy making its way to a German news magazine, *Der Spiegel* (January 16, 1995, p. 110). He simply wanted to assure his generals that even if he had to clink glasses with Western leaders, he knew who the chief enemy was.

   We don't know whether *Der Spiegel* obtained this document on its own or through the excellent German intelligence service, the BND.

49   Cox Report, Vol. I, xxxiii and 193.

50   The United States military has this combination of forces.

51   *Jiefangjun Bao* (Beijing, in Chinese), February 14, 1996. Author Nan Li has a much more detailed discussion of the PLA's interest in preemptive strikes in his "The PLA's Evolving Campaign Doctrine and Strategies," *The People's Liberation Army in the Information Age* (RAND, 1999).

52   *Zhongguo Junshi Kexue* (Beijing, in Chinese), November 20, 1997.

53   See the Report of the Defense Science Board Task Force on Information Warfare—Defense, November 1996.

54   Report of the President's Commission on Critical Infrastructure Protection, 13.

55   Ibid, A27.

56   For instance, there are clustered refineries in California, Texas, Louisiana, and New Jersey.

57   Report of the President's Commission on Critical Infrastructure Protection, A27. It's clear the commission was very worried about refineries: "Large oil refineries are also attractive targets," it repeated in Chapter 3. The Commission referred to the associated oil pipelines as "a huge, attractive, and largely unprotected target array."

58   A back door is a computer program that allows reentry at will.

59   Report of the Defense Science Board Task Force on Information Warfare, 1995, 28.
60   Cox Report, Vol. I, 11.

CHAPTER 9

1   Roilo Golez was a proud graduate of the United States Navel Academy's class of 1970. In the yearbook, his classmates identified him as "Mr. Nice Guy." He was not only a Champion Brigade boxer but also an excellent artist. His Naval Academy yearbook stated that the Philippine navy had much to look forward to. Today, all of the Philippines can be proud of his courage in going up against the Chinese navy.

2   Interview with the author (Timperlake), May 22, 1999.

3   State cable D 260845Z November 1998 (sensitive but unclassified).

4   Congressman Rohrabacher is chairman of the Subcommittee on Space and Aeronautics, where at no small political risk he took on the Loral and Hughes Corporations' improper and illegal technological assistance to the PRC. Congressman Rohrabacher's 45th District is home to a sizable population of high-tech defense workers.

5   In 1968 one of the authors (Timperlake) served on an ammunition ship, the USS *Great Sitkan*, stationed in the South China Sea. Ranging from the Philippines to Vietnam, the South China Sea is a spectacular body of water, but not as wide open as the surface map would indicate.

6   Helicopters with advanced cruise missiles can be ship killers.

7   Interview with Al Santoli, July 7, 1999.

8   An excellent example of denial occurred when the PLA navy went to significant trouble to cover up an incident at sea. The Russian navy, when it was declassifying records, reported that it had had thirty-five submarine collisions, including a June 1983 incident in which a Victor III–class nuclear-powered attack sub hit a Han-class nuclear-powered PLA navy sub. The Russian sub returned to port; the Chinese sub went to the bottom of the sea. Both sides, fearing a diplomatic incident, kept the collision quiet. The Chinese navy tried to conceal the loss of the sub by building a life-sized plywood replica. The Russians figured the PLA navy wanted it to appear the sub was still in service and not at a depth of eight hundred to a thousand meters. *Russia International Affairs*, October 19, 1995.

9   Profile of navy coastal troops, Xinhua (domestic news service, Beijing), April 22, 1999.

10   For a detailed discussion of Chinese money going into President Clinton's campaign coffers, see the authors' *Year of the Rat*.

11   *Washington Times*, August 12, 1999.

12   *Eagle Against the Sun*, by Roland H. Spector (Free Press, 1985), 34. This work is a brilliant, insightful look into the war in the Pacific. It is superbly written and researched, a must for anyone interested in World War II decision-making and combat in the Pacific theater.

13   "The Chinese Navy: Developing Blue-water Experience," by Gary Klintworth, *ASIA-PACIFIC Defense Reporter*, February–March 1998, points out that a 1997 voyage to Honolulu, San Diego, Mexico, Chile, and Peru was "the longest voyage undertaken by mainland Chinese warships in more than 500 years...."

14   *Jane's Intelligence Review*, August 1, 1995, and November 1, 1995.

15   One of the least discussed aspects of Desert Storm was the significant danger mines posed. One ship, the cruiser *Princeton* (CG-59), was cracked to the superstructure by hitting a mine, and the assault helicopter carrier *Tripoli* (LPH-10) was also damaged. Blue-water sailors have a very healthy respect for mine warfare. Due to the 100,000 mines laid by all combatants in World War II, 2,665 ships were lost. In fact, the PLA navy in a feature called "Honorable Trailblazer on the Sea" tells how the captain of minesweeper who has swept the largest number of live mines in peacetime training is often accorded a more courteous reception on the sea: "The navies the world over would follow such a convention: All types of battleships, be they large or small, should sound a siren to salute a minesweeper whenever they come across one on the sea" (*Jiefangjun Bao* [Beijing, in Chinese], December 1995).

16   In November 1995 the prestigious Naval Institute Proceedings published "A Brief Analysis of the Spratly War," a work discussing a projected naval war in the Spratlys between the United States and the PLA. In this combat, projected for the year 2006, the United States does not do well and loses "dominance in the South China Sea, at least for the time being." Commander Frank C. Borik's study "Sub Tzu and the Art of Submarine Warfare," which has the "Analysis" appended, is extremely interesting.

17   A nuclear weapon–capable Sovremenny could be a "strategic deterrent" similar to America's Trident Fleet of submarines. An attack on an American "boomer" would almost certainly start a nuclear World War III.

18   Before the advent of cruise missiles, U.S. ships had to contend with kamikazes, which the Navy first encountered in 1944 in the Philippines during the Battle of Leyte Gulf. In fact, while Vice Admiral John S. McCain (grandfather of Republican Senator John McCain of Arizona) and his Task Group One steamed to the rescue in a most impressive feat of seamanship, the kamikazes struck American carriers and began their deadly reign of terror, which lasted to the end of the war. Kamikazes damaged and destroyed a sig-

nificant number of navy surface combatants. In the June 1945 battle for Okinawa alone, 34 ships were sunk and an additional 368 were damaged. Today, the entire U.S. Navy fleet is only 323 ships. It is fair to say that cruise missiles are a much improved version of the kamikaze, and all naval planners in the world are concerned.

19   See the authors' *Year of the Rat* for the whole ugly story.

20   *Cheng Ming* (Hong Kong), May 1, 1999.

21   In early February 1995 a mutiny broke out on a North Sea Fleet Missile Corvette. The report on the incident blames drinking, gambling, and fighting as the causes. Forty officers, including the executive officer, who was the leader of the insurrection, participated in the mutiny. Twelve people were wounded; seven leading troublemakers, including the executive officer, were sentenced to death; twelve went to prison; and thirty were thrown out of the military and the Party. The executive officer may have been trying to defect, since an officer of that rank does not usually get caught up in drinking, gambling, and fighting with the crew (at least not at sea). *Cheng Ming* (Hong Kong, in Chinese), April 1, 1995.

22   See *Year of the Rat* for a full discussion of the complete sellout of national security over COSCO. Dalton and Berger will have a lot to answer for if the U.S. Navy is engaged in combat by the PRC navy or if "rogue nations" use the military equipment shipped to them in COSCO bottoms.

23   *Jane's Defense Weekly*, December 16, 1998.

24   Ibid.

25   Ibid.

26   Ibid. The PRC also has plans to develop its own carrier battle group, but this project will likely be long-term. Credible press reports reveal that the PRC is speeding up its plans for at least one aircraft carrier, but building a carrier battle group is a major undertaking that involves more than just one ship. For instance, the PRC must have a substantial force of trained naval aviators. According to speculation in the press, the PLA navy is planning a 60,000-ton, conventionally powered ship—which is smaller than an American supercarrier—with a complement of thirty-five aircraft. Some reports predict that the PLA will have an operational ship by 2006, but that is, in our opinion, optimistic.

27   Logistical support is critical. In one area, replenishment of ships at sea, if the United States is at the Pro Bowl level, the PRC is for all practical purposes playing high school ball. The critical issue is that the Chinese know their deficiencies and are rapidly working to improve, and the PLA navy is developing 10,000-ton supply vessels to support blue-water operations.

28 "Russian-Chinese Naval Cooperation to Expand," Russian Information Agency (Moscow), May 28, 1999.

29 The U.S. Navy has emerged as the world's leader in anti-submarine warfare, having developed techniques in the battle of the Atlantic and honed these skills in the past fifty years. That is why it was extremely upsetting to hear that PLA navy officers had visited the U.S. Navy's training facilities in San Diego (conversation with radio talk-show host Roger Hedgecock).

30 With the recent acquisition of Russian KILO subs, the PRC has moved ahead rather effectively. The KILO, a diesel submarine, is known as the "black hole" because of its stealth characteristics. As U.S. Navy Captain Jim Patton pointed out in a special report in *International Defense Review* (November 1995), "The difference between diesel-electric and SSNs [nuclear-attack boats] is not that the latter are faster, bigger, deeper diving, or more expensive, but that the former are defensive, and the latter offensive weapons systems. Indeed, if an emerging navy desired the most cost-effective, one-ship fleet to defend its home waters, that single ship would have to be a modern diesel-electric submarine."

31 "Exploring the Secrets of China's Frogman Unit," *Zou Bo Beijing Guofang*, February 15, 1999.

32 Randy "Duke" Cunningham and his radar intercept officer, Bill "Irish" Driscoll, made aviation history in Vietnam by becoming the first aces (five confirmed kills). One day, May 10, 1972, a "Turkey Shoot" took place— eight MiGs fell to the navy, and another three to the air force. Cunningham and Driscoll were credited with three kills in one engagement, including the superb North Vietnamese fighter pilot "Colonel Tomb." Cunningham and Driscoll's plane was hit after they killed Colonel Tomb; they were barely rescued by marine helicopters. In all, Representative Cunningham flew three hundred combat missions and was awarded the Navy Cross, two Silver Stars, fifteen Air Medals, and the Purple Heart.

33 Personal interview with Representative Cunningham.

34 The PRC's fighter assets include the Su-27, an aircraft for which current fighter pilots still hold tremendous respect. In addition, credible press reports reveal that the PLA is spending $2 billion for sixty Su-30 aircraft. The Su-30 is a very capable aircraft with advanced weapons systems; it would be a handful for any American fighter pilot.

   If the United States is to be prepared to counter the PLA air force, the F-22 should, in the opinion of the authors, be developed immediately.

35 Allowing Chinese pilots to visit Top Gun is a small part of the military-to-military exchange conducted under the Clinton-Gore administration. PLA

troops have observed American anti-submarine forces, logistical teams, marines at Quantico, and more. The Chinese seem to be gaining an edge in this exchange. Noted military expert Dr. Michael Pillsbury observed, "China is not as friendly to the Pentagon as the Pentagon is to China" (American Enterprise Institute Conference on the PLA, September 15, 1997).

CHAPTER 10

1   *Washington Times*, July 30, 1999.

2   "Republic of China" and "People's Republic of China" are confusing to some. When we use "Taiwan," we mean the Republic of China on Taiwan.

3   *Star* (Malaysia), August 12, 1999.

4   Chen, 99.

5   Politburo member Zou Ziahua, now retired, is married to the daughter of Marshal Yeh Jianying and was chairman of Norinco, China's largest arms producer.

6   After the United States dropped diplomatic recognition of the Republic of China in 1979, the chief U.S. official on Taiwan was called the head of the American Institute in Taiwan. The current representative, Lynn Pascoe, told a group of visiting congressional staff members, including the author (Triplett), that the Vietnamese had "defeated the American military." This did not go over well with the two Purple Heart winners in the staff delegation.

7   See, generally, the authors' *Year of the Rat.*

8   The Chinese Communists didn't want Lee welcomed to America because they felt that if the United States recognized a Taiwanese official, it would violate Beijing's "One China" policy, by which the PRC considers Taiwan a renegade province. But Beijing's One China policy overlooks the fact that the Republic of China has been in existence since 1911, and that the entire Taiwanese government has been elected by the people, whereas the PRC has never held a popular election.

9   The lone dissenter was former Senator Bennett Johnston (D-Louisiana), who left Congress and later sought "to profit from business ventures in China" (Mann, 322). Johnston appeared as a guest at the White House state dinner for Premier Zhu Rongji.

10 One myth holds that Taiwan somehow "bought" support for the Lee visit, but congressional opposition to Beijing was so strong that those kinds of overwhelming votes were inevitable.

11 The Scud missiles fired in the Gulf War were not nuclear capable at the time.

12 We believe the best analysis of the 1995 and 1996 PLA missile bullying of Taiwan is "China's Missile Tests Show More Muscle," by Richard Fisher and Greg Gerardi, *Jane's Intelligence Review*, March 1997, 125.

13 Private estimate of a distinguished European economist.

14 This private comment illustrates the substantial percentage of Taiwan's elite who have American connections.

15 Floor speech, United States Senate, August 11, 1995 (see *Congressional Record*, S12417).

16 *New York Times*, January 24, 1996. There may have been a second nuclear threat in this same time period. Mr. Wang Chi, a Congressional Research Service employee known to be close to Beijing's leadership, was quoted in a Singapore newspaper claiming security boss Qiao Shi had threatened to incinerate New York with nuclear weapons. In a subsequent story, Wang claimed to have been misquoted. See the *Straits Times* (Singapore), January 17, 1996, and January 19, 1996.

17 Cox Report, Vol. I, 84.

18 Ibid.

19 *Los Angeles Times*, April 30, 1999.

20 During this tense time period the authors were receiving "Do something!" telephone calls and e-mail from various parts of the U.S. national security apparatus. Appeals to Capitol Hill are fairly common when an administration goes into paralysis.

21 The best unclassified blockade scenario is offered in Chapter 8 of *The Coming Conflict With China,* by Richard Bernstein and Ross H. Munro (Knopf, 1997).

22 We take responsibility for what follows, but we also acknowledge a debt of gratitude to U.S. Air Force Major Mark A. Stokes.

23 "Report to Congress Pursuant to Section 1226 of the FY 98 National Defense Authorization Act."

24 *Jiefangjun Bao* (Beijing, in Chinese), February 14, 1996.

25 See "Dialectics of Defeating the Superior with the Inferior," by Senior Colonel Shen Kuiguan, reprinted in *Chinese Views of Future Warfare*, edited by Dr. Michael Pillsbury (National Defense University Press, 1996), 213.

26 Title of article in *Jiefangjun Bao* (Beijing, in Chinese), June 2, 1998.

27 *Mao Tse-tung on Guerrilla Warfare*, translated by Samuel B. Griffith (Praeger, 1963).

28 "The Security Situation in the Taiwan Strait," a report to Congress, February 1999.

29   See *Betrayal*, by Bill Gertz (Regnery, 1999), 105.

30   Cox Report, Vol. I, 191.

31   Ibid, xvii.

32   "The Security Situation in the Taiwan Strait."

33   *Flight International*, August 31, 1995.

34   "Denial and Deception," known as "D&D," is very controversial among American PLA-watchers. Some observers reject the idea altogether. Other observers, including the authors, believe that the PRC practices D&D across the board in the defense area. If so, the PLA has substantially higher capabilities than thought according to the current, politically correct, analysis. Unfortunately, extensive discussion of this issue cannot realistically take place in an unclassified environment.

35   *Sun Tzu, The Art of War*, 66.

36   "Future Military Capabilities and Strategy of the People's Republic of China," Report to Congress Pursuant to Section 1226 of the FY1998 National Defense Authorization Act.

37   Cox Report, Vol. I, xiii.

38   "Investigation of Illegal or Improper Activities in Connection with 1996 Federal Election Campaigns," Senate Committee on Governmental Affairs, 1767.

39   Peter Lee, Wen-ho Lee, and an unidentified person formerly at Lawrence Livermore Labs.

40   The Cox Committee believes that scientist Peter Lee was co-opted through appeals to "his ego, his ethnicity, and his sense of self-importance as a scientist" (Cox Report, Vol. I, 90).

41   *China Times*, August 17, 1999.

42   *The Haunted Wood: Soviet Espionage in America—The Stalin Era*, by Allen Weinstein (Random House, 1999).

43   AFP, April 28, 1999; also, *Taipei Tzu-Li Wan Pao* (in Chinese), April 28, 1999.

44   22 US Code 3302, enacted April 10, 1979.

45   This description is based on the author's (Triplett's) March 1997 visit aboard a Sovremenny-class destroyer in the Middle East, as well as the account in *Jane's Fighting Ships 1993–1994*.

46   The Hiroshima bomb had an explosive power of 12.5 kilotons.

47   There are unconfirmed reports that Russia has sold additional SS-N-22s to China to retrofit existing PLA navy vessels.

48   Total explosive power: 1.6 megatons, 125 times the power of the Hiroshima bomb. The SS-N-22 also comes in a high-explosive version, which can sink an

aircraft carrier or Aegis cruiser. But, given the corruption endemic to Russia and the lingering anti-American feeling, there is no guarantee that the PLA navy could not obtain nuclear warheads. A 1996 CIA report called the Russian military "demoralized and corrupted" and warned that Russian civilian leaders might lose control of the nuclear arsenal to the military (*Washington Times*, June 15, 1999). Worse, now that the PLA has tested its own stolen version of the American W-88 nuclear warhead, there would be nothing to stop it from adapting the W-88 to the Russian missile. Finding renegade Russian scientists to help with the adaptation should not be difficult.

The Sovremennys are far and away the most expensive ships the PLA navy has ever purchased abroad. The authors believe that Beijing would not put up this amount of foreign exchange for non-nuclear weapons. One way or another, we believe the PLA navy's SS-N-22s will be nuclear.

49 The authors are aware that U.S. Navy's attack submarine fleet could quickly take out a PLA navy Sovremenny; the question is who gets off the first shot.

50 Admiral McVadon was formerly the U.S. defense attaché in Beijing (*South China Morning Post*, June 7, 1998).

51 This comes from a book published in Taipei in 1995.

52 The authors have based their conclusion on a conversation with two experts on the Taiwanese military.

53 *Taipei Hsin Hsin Wen* (in Chinese), March 18, 1999.

54 *Taipei Hsin Hsin Wen* (in Chinese), March 31, 1999.

55 Dr. Lin made his prediction in the *International Defense Review* issue of February 1995, five months *before* the first missile test off Taiwan's shores.

56 See "Nuclear Weapons Viewed as a War Deterrent" in *Taipei Ch'uan-ch'iu Fang-wei Tsa* (in Chinese), March 15, 1999.

57 There are many more possibilities than what is presented here. This just scratches the surface of what can be done.

CHAPTER 11

1 Korean Central Broadcasting Network (Pyongyang, in Korean), February 4, 1999.

2 North Korea is a nation so ravaged by famine that, according to credible reports, some Koreans have resorted to cannibalism in order to survive.

3 Kyodo, July 11, 1997. The agreement was apparently signed in May 1996.

4 For an example of Xiong's contact with the North Koreans, see Xinhua, July 13, 1999.

5 *Wall Street Journal*, March 15, 1994.

6 *The Military Balance 1998/1999*, 178.

7   North Korea has an extensive biological warfare program, perhaps created
     with Cuban assistance. See *Jane's Intelligence Review*, August 1998.

8   *Washington Times*, July 21, 1999.

9   Ibid.

10   Yonhap News Agency (Seoul), as reported by the Kyodo News Service, September 12, 1995.

11   Unpublished manuscript, "Bombs from Beijing," Wisconsin Project on
     Nuclear Arms Control, May 1991, 18. Milhollin is a professor at the University of Wisconsin.

12   *Cruise Missile Proliferation in the 1990s*, by W. Seth Carus (Praeger, 1992).

13   Cox-Dicks press conference transcript, May 25, 1999.

14   *Washington Times*, February 23, 1999.

15   *Washington Post*, December 25, 1987.

16   *New York Times*, February 23, 1989.

17   *Wall Street Journal*, January 17, 1994.

18   Kyodo News Service, July 11, 1997.

19   *Times of India*, July 18, 1999.

20   *Korea Times*, April 2, 1999.

21   Interview with the author (Triplett), in Washington, 1985.

22   Interview with the author (Triplett), in Beijing, 1993.

23   Korean Central News Agency (Pyongyang), August 3, 1998.

24   Xinhua, September 1, 1998.

25   Kyodo, December 1, 1997. We don't know if Xiong found it amusing to trick
     the Liberal Democratic Party leadership.

26   Associated Press, March 5, 1999.

27   AFP, March 26, 1999.

28   *Washington Post*, August 18, 1998.

29   *Seoul News Plus* (in Korean), July 14, 1999.

30   *Defense of Japan 1999* (Japan Defense Agency, 1999), as reported by the
     Associated Press on July 27, 1999.

31   Cox Report, Vol. I, 68.

32   The author (Triplett) attended the dinner.

33   "China's Ballistic Missile Programs," by Hua Di and John Wilson Lewis,
     International Security, Fall 1992.

34   Pacific News Service, May 27, 1998.

35   See Cox Report, Vol. I, 180; also, see any edition of *The Military Balance*.

36   *Scotsman*, March 12, 1999.

37   *Wen Wei Po* and *Zhongguo Tongxun She* (both in Chinese), August 3, 1999.

38   Cox Report, Vol. I, 180.

39   These islands are known as the "Senkaku" in Japanese, "Diaoyutai" in Chinese.

40   *Defense News*, August 28, 1995.

41   Ibid.

42   Speech before the Center for Strategic and International Studies, Washington, June 29, 1999.

43   The authors recognize that there always low-tech solutions, some more effective than others. Major General Les Bray, a World War II pilot who flew in China, Burma, and India, told the author (Timperlake) that the Americans in that theater worked around their lack of high-tech equipment. For instance, when radar was weak or nonexistent, thousands of troops would look out for Japanese war planes and warn of any approach by phone or radio. It wasn't a perfect system, but it was still an attempt to win the war.

44   "Sun Tzu Art of War in Information Warfare: Information Terrorism: Can You Trust Your Toaster?" by Matthew G. Devost, Brian K. Houghton, and Neal A. Pollard, Institute for National Strategic Studies, September 1998.

45   *Strategic Information Warfare* (RAND, 1996), 65.

46   See <http://www.worldwideminerals.com> and <http://www.renewables.ca>.

CHAPTER 12

1   AFP, May 4, 1998.

2   *Times of India*, October 5, 1994.

3   Moviegoers will recall the dramatic point in the film *The Hunt for Red October*, based on the Tom Clancy novel, when the Soviet SSBN races full throttle through an underwater canyon. That's not possible without meticulous mapping of the ocean floor done years in advance.

4   Lintner, "Myanmar's Chinese Connection," 23.

5   Democratic Voice of Burma (anti-SLORC radio), June 20, 1998.

6   Xinhua, November 25, 1998.

7   Xinhua, May 5, 1999.

8   Reuters, February 21, 1997, and April 4, 1997.

9   See, for example, Xinhua, November 19, 1998, and December 12, 1998.

10   Telephone interview with the author (Triplett), October 14, 1996.

11   Letter to Mr. Haruna dated July 15, 1995, from the United States Army Intelligence and Security Command, Fort George G. Meade, Maryland.

12   U.S. Army intelligence report, dated May 14, 1975, portions declassified and attached to the letter to Mr. Haruna.

13   Interview with Indian defense officials, September 1995, in New Delhi.

14   U.S. Army intelligence report, dated April 20, 1983, distributed by the U.S. Joint Chiefs of Staff, declassified and attached to the letter to Mr. Haruna.

15   See *Washington Times*, February 5, 1996; *Washington Post*, February 7, 1996; *Nucleonics Week*, February 8, 1996.

16   Senate Hearing 104-510, February 22, 1996, 35.

17   Telephone interview with Mark Hibbs, February 8, 1996.

18   Statement by a State Department spokesperson at a news conference in Washington, D.C., May 10, 1996.

19   Answers by Assistant Secretary of State Winston Lord to questions submitted by Senator Robert Bennett (R-Utah), August 6, 1996.

20   *Washington Post*, June 13, 1996.

21   The Manhattan Project was the multibillion-dollar project to develop the atomic bomb during World War II.

22   *Periscope Daily Defense News Capsules*, April 2, 1996. See also "Pakistan Persists with Nuclear Procurement," by Andrew Koch, *Jane's Intelligence Review*, March 1997.

23   *Washington Times*, April 11, 1996.

24   *Washington Times*, October 9, 1996.

25   Ibid.

26   Ibid.

27   *Washington Times*, April 6, 1991; *Washington Post*, April 6, 1991.

28   "On the Nuclear Edge," by Seymour Hersh, *The New Yorker*, March 29, 1993.

29   Burrows and Windrem, 82.

30   TELs are used to move missiles around.

31   *Time*, April 22, 1991.

32   Statement by Margaret Tutwiler, spokesperson for the Department of State, February 21, 1992.

33   Ibid.

34   *Defense News*, April 6, 1992.

35   Associated Press, May 14, 1992.

36   *Washington Post*, December 4, 1992; *Los Angeles Times*, December 4, 1992.

37   *Independent*, December 8, 1992.

38   *New York Times*, May 6, 1993.

39   Statement by Michael McCurry, spokesman for the Department of State, August 24, 1993.

40   "Joint Statement of the United States of America and the People's Republic of China on Missile Proliferation," Washington, October 4, 1994.

41 *New York Times*, October 5, 1994.

42 *Arms Control Today*, December 1994.

43 *Washington Post*, July 3, 1995.

44 Senate Hearing 104-510, February 22, 1996, 35.

45 *Far Eastern Economic Review*, October 3, 1996.

46 *Far Eastern Economic Review*, July 18, 1991.

47 *Washington Times*, June 21, 1996.

48 *Washington Times*, June 12, 1996.

49 *Washington Post*, August 25, 1996. See also *Time*, June 30, 1997.

50 Kyodo News Service, July 3, 1997.

51 U.S. foreign service officer who wishes to remain anonymous. Telephone interview with the author (Triplett), October 3, 1996.

52 James A. Baker III, *The Politics of Diplomacy* (Putnam, 1995), 593.

53 The *Los Angeles Times* (May 23, 1996) revealed that Poly has been the marketing agent for the M-11 missile to Pakistan. Lieutenant Colonel Liu Chaoying, daughter of PLA General Liu Huaqing, was also part of the deal.

54 The delegation included the author (Triplett). The Pressler Amendment, passed by Congress in 1986, was named for its chief sponsor, former Senator Larry Pressler (R-South Dakota).

55 The authors do not know whether there is any connection to possible campaign contributions from American defense contractors.

CHAPTER 13

1 Since the publication of our book *Year of the Rat*, we have discovered that a growing, bipartisan coalition of Americans is recognizing the threat to America's security. We have spoken to countless men and women, Democrats and Republicans, who now see the consequences of President Clinton's betrayal of his duty to protect America's national security.

APPENDIX A
THE PLA IN FACTS AND FIGURES*

TOTAL ARMED FORCES

The Chinese People's Liberation Army (PLA) has about 3 million troops on active duty, plus an additional million reserves. In contrast, the United States has about half the active-duty forces (1.4 million), plus 1.3 million reservists.

OVERALL MILITARY COMMAND

The People's Republic of China's (PRC's) armed forces are under the command of the Chinese Communist Party's (CCP's) seven-member Central Military Commission. The chairman of the Central Military Commission is CCP General Secretary Jiang Zemin. Jiang has no military experience, although he spent part of his career in defense industries. The other six members of the Central Military Commission are all senior generals.

Under the Central Military Commission are four major departments: the General Staff Department, the General Political Department, the General Logistics Department, and the General Armaments Department. The military services—army,

*Some of this material comes from *The Military Balance 1998/1999,* by the International Institute for Strategic Studies (Oxford, 1998). To the extent possible, statistics are good as of the summer of 1999.

navy, air force, and Second Artillery (rocket forces)—and the military regions report here. The defense minister in the Chinese system is a significant figure, but the Ministry itself is not as important.

STRATEGIC MISSILE FORCES

The strategic (nuclear) rocket forces of the Second Artillery are thought to have more than twenty intercontinental ballistic missiles—East Wind Models 4 and 5. Some of these missiles have multiple warheads, and some are targeted on the United States. About fifty intermediate-range ballistic missiles—East Wind Models 3 and 21—are targeted on neighboring countries, including Japan and India. The East Wind Models 3, 4, and 5 use 1950s-era technology and are scheduled to be replaced by solid-fuel, mobile missiles—the East Wind Models 31 and 41— in the next few years. The Second Artillery also has an indeterminate number of short-range ballistic missiles—East Wind Models 11 and 15—a substantial number of which are targeted at Taiwan.

The PLA navy is thought to have one nuclear-powered ballistic missile submarine armed with twelve intermediate-range missiles. In the early 1980s the sub successfully launched a missile while submerged, but this boat has had technical trouble and has made few cruises, none of them far from the Chinese coast. A new nuclear missile for submarines is nearing the end of its development stage and should be deployed in the next few years, which may mean new nuclear subs, probably based on Russian designs.

THE ARMY

China is divided into seven military regions named for cities (e.g., "Beijing Military Region"), and the military region commanders and political commissars are important figures. The military regions contain a number of military districts and garrison commands whose boundaries follow city or provincial lines.

Since the mid-1980s the PLA army has been organized into twenty-four Group Armies of about 60,000 men each. Most Group Armies have three infantry divisions, one tank brigade, one artillery brigade, and one air defense brigade, although some have an armor division (10,000 men) rather than a brigade (approximately 2,500 men). The PLA is attempting to form "rapid reaction" forces that can move anywhere in the country within twenty-four hours. This is an outgrowth of the PLA's Tiananmen experience, but there is increasing evidence that these forces may be training for duty in Taiwan.

The army has mostly 1960s vintage tanks, artillery, and infantry weapons—strong and serviceable. Before he was assassinated in Europe, Gerald Bull brought his modern artillery designs to China, but it is uncertain how many of them have been produced in China. Most of the PLA's anti-tank missiles are copies of Western models. The PLA has substantially fewer helicopters than other military forces its size deploy.

THE NAVY

In addition to the single nuclear-powered ballistic missile submarine, the PLA navy has five nuclear attack submarines, some of which are being armed with cruise missiles. The PLA navy has purchased four very capable KILO-class modern diesel submarines from Russia. It has embarked on a domestic submarine-building program with Russian technical assistance. Its remaining submarine fleet is composed of 1950s Soviet-designed diesel boats. In total, the Chinese have about sixty subs they could put to sea.

The PLA navy will undergo a major change when the two Russian-built Sovremenny-class destroyers arrive. In Russian service, the Sovremenny is armed with eight supersonic nuclear-tipped anti-shipping missiles. The PLA navy is reported to have purchased fifty missiles along with the Sovremenny destroyers, and some of these may be fitted onto existing PLA ships.

The PLA navy has a little more than 50 home-built major surface warships (destroyers and frigates) armed with cruise missiles. It also has more than 150 missile craft for launching cruise missiles. The Chinese have exported such vessels to Iran, and Pakistan has used the design to build its own ships. The PLA navy has more than 70 amphibious ships of various kinds that could move about one division to Taiwan in the event of hostilities.

A substantial naval air arm is armed with Chinese copies of 1950s-era Soviet bombers and fighters. The bombers are armed with cruise missiles.

THE AIR FORCE

The PLA air force has approximately 3,500 combat aircraft, the vast majority of which are Chinese-built copies of 1950s-era Soviet designs. In 1992 the PLA air force took delivery of its first Russian Su-27s, a modern third-generation fighter aircraft. At present, the PRC is assembling variants of the Su-27 under contract with Russia.

The PLA air force is also working with Israel to develop its own fighter aircraft based on the Lavi, which is itself based on the American F-16. According to one unconfirmed report, the PLA air force received an American F-16 from Pakistan, presumably as part of another arms deal. The prototype of this Chinese-Israeli hybrid first flew in 1998.

By 2005 the PLA air force is expected to have five hundred modern, all-weather fighters in its inventory.

The PLA air force is thought to be deficient in transports, but it has recently acquired ten IL-76 Candids from Russia. China's civilian aircraft fleet has grown considerably in recent years, and these aircraft would be available in case of hostilities.

In Chinese practice, airborne forces—paratroops—are part of the air force. The PLA air force has one airborne army composed of three divisions that is stationed in the center of the country, ready for any emergency. These troops are principally designed to support the People's Armed Police, but they have been observed training for a possible airborne assault on Taiwan.

PEOPLE'S ARMED POLICE

The People's Armed Police is a paramilitary group close to one million strong. Many of its forces are ex-PLA officers and soldiers in police uniforms. Since the Tiananmen massacre in 1989, the People's Armed Police has greatly expanded and has received new training and equipment. Although it has some incidental border defense duties, its prime role is to suppress democracy in China.

BUDGET

The defense budget of the PRC announced in 1999 is $12.6 billion. But that figure does not include many items that would normally be included in most other countries' defense budgets, such as weapons procurement and research and development. Since the Tiananmen massacre, defense spending has increased considerably. Even Beijing's announced budget figures are misleading, since PLA salaries are quite low compared to those of Western armed forces. According to noted economist Charles

Wolf of the RAND Corporation, Beijing's defense budget is equivalent to about $140 billion, in Western terms.

DEFENSE INDUSTRIAL BASE

The PRC's defense industrial base had its modern beginnings 150 years ago with the "strengthening movement" of the Qing (Ching) Dynasty. During their occupation of Manchuria, the Japanese built munitions factories that the PRC inherited in 1945. The biggest boost came in the 1950s, when there were close working relations between Stalin's Soviet Union and Mao's China. Many of the PRC's top civilian leaders were trained in Moscow in the 1950s, and a number of the PRC's defense industries are still based on Soviet models.

By the 1960s Moscow and Beijing had fallen out, and Mao believed the West would attack the PRC from the eastern seaboard and the Soviet Union would invade from the north. In response, the PRC poured billions of dollars into moving and duplicating defense industries to the most remote parts of central China, called the "Third Line." As a result, the PRC can produce far more basic arms than either the PLA or the world arms bazaar can absorb. For the past fifteen years the PRC has been trying to deal with these bloated defense plants. There has been a lot of talk of "defense conversion," most of which turns out to be "defense diversification"—that is, civilian product lines are set up beside the existing military product lines and are not true substitutes for them.

The Chinese government is trying to encourage "pockets of excellence" among the defense industries; missiles and defense electronics are getting the most attention. It helps that PRC President Jiang Zemin was once the minister of the electronics industry. In many cases, the government has actively encouraged tie-ins with foreign companies who have militarily useful technology to transfer.

SCIENCE AND TECHNOLOGY

The PRC's first big science and technology boost came from two sources—Chinese students returning from the United States and a younger generation trained in Moscow institutes. This group produced the PLA's nuclear weapons and missile systems. The PLA is now actively recruiting some of the Chinese students sent to the West and Japan in the past twenty years. The extent of this recruiting is kept highly confidential in China. There is also a companion program of aggressive espionage, of which the recently revealed Los Alamos spying is only part.

APPENDIX B
PERENNIAL AGGRESSION:
A PLA CHRONOLOGY

- *October 1, 1949:* The PLA imposes Communism on China; fifty million people die from execution, torture, and starvation.
- *June 1950:* Former PLA soldiers in North Korean uniforms launch a surprise attack on South Korea.
- *Fall 1950:* PLA "volunteers" enter the Korean War, killing two million civilians.
- *Fall 1950:* The PLA conducts a surprise attack on Tibet.
- *1950s–1970s:* PLA associates conduct armed subversion in the nations of Southeast Asia.
- *Fall 1962:* The PLA launches a surprise attack on India.
- *Christmas Day, 1968:* Thousands of PLA "volunteers" stream across the Burmese border.
- *March 1969:* PLA troops clash with Soviet forces in the first direct fighting between two declared nuclear powers.
- *February 1979:* The PLA hits Vietnam with a surprise attack.
- *1970s–1990s:* PLA intelligence penetrates defense laboratories in the United States, stealing vital American secrets.
- *1980s–1990s:* PLA arms smugglers become the world's leading proliferators of weapons of mass destruction to terrorist regimes.

- *1988:* The PLA becomes the protector of the Burmese military junta, which has seized power from Burma's democratically elected leadership.
- *June 4, 1989:* The PLA massacres thousands of innocent Chinese in and around Beijing's Tiananmen Square.
- *1990:* A grateful Chinese Communist Party rewards the PLA with money and other favors in the wake of the Tiananmen Square massacre; the military buildup truly begins.
- *Spring 1992–present:* PLA intelligence uses illegal campaign contributions as an instrument to penetrate the Clinton-Gore administration.
- *February 1995:* The Philippines discovers that the PLA has taken over Philippine territory in the South China Sea.
- *Summer 1995:* The PLA's Second Artillery fires nuclear-capable missiles off the shores of democratic Taiwan.
- *Spring 1996:* The PLA acquires missile technology from two American corporations, Hughes and Loral.
- *Spring 1996:* The PLA's Second Artillery fires a second round of missiles toward Taiwan in an effort to scuttle democratic elections on the island.
- *June 1997:* United States Secretary of Defense William Cohen complains of the threat to American forces in the Persian Gulf from illicit sales of Chinese cruise missiles to Iran.
- *1998–1999:* The PLA helps the Chinese Communist Party suppress the democratic movement in China.
- *February 1999:* The PLA establishes an electronic surveillance post in Cuba.
- *May 1999:* The Cox-Dicks Committee accuses the Chinese Communists of using espionage to steal neutron

bomb technology and other strategic weapons systems from the United States.

■ *Summer 1999:* The PLA steps up its bullying and intimidation of Taiwan.

■ *Summer 1999:* The PLA test-fires a new intermediate-range ballistic missile based in part on stolen or acquired American technology.

INDEX

ACDA. *See* Arms Control and Disarmament Agency
Africa, 134
Aikman, David, 26–27, 38
airport radar signals, 122
Albright, Madeleine, 102, 109–110, 115, 174–175
Algeria: Iraqi advanced weapons programs and, 91; nuclear weapons site in, 85–88; plutonium production in, 95
Alibek, Ken, 115
America. *See* United States
American Strategic Air Command, 165
Amnesty International, 28–29
Appomattox, 134
Argentina, 88
arms companies: auxiliary companies and, 74; Chinese, 72–73; PLA-associated, 74; PLA-owned, 73–74; profits from, 78–80; sixteen-character policy and, 75–77
Arms Control and Disarmament Agency (ACDA), 114, 115
Armscor, 94
"Arms Transfers to State Sponsors of Terrorism," 98

Army War College's Strategic Studies Institute, 126
"The Art of War" (Sun Tzu), 127
Artisu Corporation, 176
Asia: Communism in, 60; Communist China and, 83, 133; PLA devastation in, 134; PLA threat to, 14
Associated Press, 24
Atlas Mountains, 85
Atomic Energy Organization, 100
Aung San Suu Kyi, 81
Australia, 133

Baker, James, 72, 112, 113, 191, 194
Ball, Desmond, 128
ballistic missile submarines (SSBNs), 147, 185
Bangladesh, 186
Bartholomew, Reginald, 71–72
Baxter, Jeff "Skunk," 139
Beijing: under martial law, 45; Tiananmen Square massacre and, 21, 22, 24, 26–27. *See also* Communist China
Beijing Workers Autonomous Union, 30
Bennett, Robert, 103, 188–189

255